New Perspectives in Micro-Nano Optical Design and Manufacturing

New Perspectives in Micro-Nano Optical Design and Manufacturing

Guest Editors

Xing Peng
Shiqing Li
Xiang Sun

MDPI

Basel • Beijing • Wuhan • Barcelona • Belgrade • Novi Sad • Cluj • Manchester

Guest Editors
Xing Peng
College of Intelligence Science and Technology
National University of Defense Technology
Changsha
China

Shiqing Li
School of Physics
Zhejiang University of Technology
Hangzhou
China

Xiang Sun
School of Electrical and Automation Engineering
East China Jiaotong University
Nanchang
China

Editorial Office
MDPI AG
Grosspeteranlage 5
4052 Basel, Switzerland

This is a reprint of the Special Issue, published open access by the journal *Photonics* (ISSN 2304-6732), freely accessible at: https://www.mdpi.com/journal/photonics/special_issues/5R86914YGU.

For citation purposes, cite each article independently as indicated on the article page online and as indicated below:

Lastname, A.A.; Lastname, B.B. Article Title. *Journal Name* **Year**, *Volume Number*, Page Range.

ISBN 978-3-7258-8123-9 (Hbk)
ISBN 978-3-7258-8124-6 (PDF)
https://doi.org/10.3390/books978-3-7258-8124-6

Contents

About the Editors

Xing Peng

Xing Peng, Ph.D., is a Professional Technical Assistant Researcher at the College of Intelligence Science and Technology, National University of Defense Technology. He is a Deya Young Scholar, recipient of the National Outstanding Doctoral Dissertation Nomination, Shanghai Outstanding Doctoral Dissertation, and Fudan Outstanding Doctoral Dissertation. He is also a Senior Member of the Chinese Society for Optical Engineering. His current research interests include intelligent defect detection and damage assessment, optical metasurface devices, and laser-induced damage threshold enhancement. He has led seven projects, including the National Natural Science Foundation Youth Program and the National University of Defense Technology Independent Innovation "Golden Idea" Program, and has participated as a key member in eight projects, including the National Major Scientific Instruments Program. He has received honors such as the National Defense Science and Technology Commendation, the Science and Technology Progress Award from the Chinese Society for Optical Engineering, the Innovation Paper Nomination Award from the Chinese Society for Optical Engineering, the Ruijun Innovation Award, the Best Paper Award (Impact Factor: 16.1), and the Shanghai Outstanding Doctoral Dissertation. He has published over 50 SCI/EI papers in journals such as the top journal in the global manufacturing field, *IJEM*, filed more than 60 patents, and obtained six software copyrights. He serves as a Youth Editorial Board Member and Guest Editor for several SCI journals and is a reviewer for over ten SCI journals. He has guided more than 30 students in seven national/provincial innovation and entrepreneurship training programs, receiving three awards in national/provincial competitions, and has been honored as an Outstanding Instructor in Hunan Province.

Shiqing Li

Shiqing Li, PhD, Associate Professor (University-appointed), received his PhD degree from the Department of Optical Science and Engineering, School of Information Science and Technology, Fudan University in September 2020, and subsequently joined the College of Physics, Zhejiang University of Technology. His current research interests include electromagnetic metamaterials and metasurfaces, surface plasmons, unidirectional waveguides, etc. He has published more than 20 papers in internationally renowned journals such as *Nature Communications*, *Advanced Science*, *Nanophotonics*, *Optics Letters*, *Optics Express*, and *Applied Physics Letters* as the first author or corresponding author. He is currently leading one project supported by the National Natural Science Foundation of China for Young Scholars.

Xiang Sun

Xiang Sun, Ph.D., is an Associate Professor and Master's Supervisor at the School of Electrical and Automation Engineering at East China Jiaotong University. He received his Ph.D. degree in Optical Engineering from Fudan University in 2021. From 2022 to 2024, he was a postdoctoral fellow in the Electronic Science and Technology program at Fudan University. His research primarily focuses on fringe projection profilometry and optical freeform surface lenses. He has led projects including the Young Scientists Fund of the National Natural Science Foundation of China, several projects funded by the Jiangxi Provincial Natural Science Foundation and the Jiangxi Provincial Department of Education, as well as multiple industry-commissioned projects. To date, he has published over 30 SCI-indexed papers, with representative achievements featured in journals in the field such as *Optics Express*, *Optics*

& Laser Technology, and *Lighting Research & Technology*, etc. He has served as a long-term reviewer for prestigious journals including *IEEE Transactions on Industrial Informatics*, *Optics & Laser Technology*, *Optics Express*, etc. Additionally, he holds 12 authorized patents and 1 software copyright. In 2023, he was honored with the Third Prize within the Science and Technology Progress Award from the Chinese Society for Optical Engineering and the Second Prize within the Science and Technology Progress Award from Jiangxi Province. Dr. Sun also serves as a review expert for the National Natural Science Foundation of China, a dissertation review expert for the Academic Degrees Committee of the Ministry of Education, a member of the Vision Inspection Committee of the China Society of Image and Graphics, an evaluation expert for the Jiangxi Provincial Department of Industry and Information Technology, and a Guest Editor for the SCI journal *Photonics*.

photonics

MDPI

Article

Defect Detection Method for Large-Curvature and Highly Reflective Surfaces Based on Polarization Imaging and Improved YOLOv11

Zeyu Yu [1,2], Dongyun Wang [1,2] and Hanyang Wu [1,2,*]

1 School of Engineering, Zhejiang Normal University, Jinhua 321005, China
2 Key Laboratory of Intelligent Operation and Maintenance Technology and Equipment for Urban Rail Transit in Zhejiang Province, Jinhua 321005, China
* Correspondence: wuhanyang@zjnu.edu.cn

Abstract: In industrial manufacturing, product quality is of paramount importance, as surface defects not only compromise product appearance but may also lead to functional failures, resulting in substantial economic losses. Detecting defects on complex surfaces remains a significant challenge due to the variability of defect characteristics, interference from specular reflections, and imaging non-uniformity. Traditional computer vision algorithms often fall short in addressing these challenges, particularly for defects on highly reflective curved surfaces such as aircraft engine blades, bearing surfaces, or vacuum flasks. Although various optical imaging techniques and advanced detection algorithms have been explored, existing approaches still face limitations, including high system complexity, elevated costs, and insufficient capability to detect defects with diverse morphologies. To address these limitations, this study proposes an innovative approach that analyzes the propagation of light on complex surfaces and constructs a polarization imaging system to eliminate glare interference. This imaging technique not only effectively suppresses glare but also enhances image uniformity and reduces noise levels. Moreover, to tackle the challenges posed by the diverse morphology of defects and the limited generalization ability of conventional algorithms, this study introduces a novel multi-scale edge information selection module and a Focal Modulation module based on the YOLOv11 architecture. These enhancements significantly improve the model's generalization capability across different defect types. Experimental results show that, compared to state-of-the-art object detection models, the proposed model achieves a 3.9% increase in precision over the best-performing baseline, along with notable improvements in recall, mAP50, and other key performance indicators.

Keywords: polarization imaging; defect detection; YOLOv11; deep learning

1. Introduction

In industrial production, defects often arise on product surfaces due to issues in manufacturing processes and techniques. In severe cases, these defects can result in product failure and significant economic losses. Currently, surface defect detection methods include laser inspection [1], eddy current testing [2], infrared detection [3], and machine vision-based detection [4]. Among these, machine vision-based detection is the most widely applied and has been extensively used for detecting defects on steel surfaces [5], integrated circuits [6], and cylindrical batteries [7]. However, detecting defects on complex curved surfaces using vision-based methods still faces several challenges. First, specular reflection on smooth surfaces can cause glare interference, causing overexposure and distortion in certain image regions, which results in the loss of defect information in those areas. Second, the morphology of surface defects varies depending on their location on the curved surface, leading to instability in defect shapes [8]. Additionally, surface defects are often small in scale [9], exhibit inconspicuous features, and have diverse morphologies, making them difficult to extract. Traditional vision algorithms, such as edge detection [10], filtering methods [11], and texture-based algorithms [12], lack the generalization ability to handle defects of varying scales and forms. When surface defects such as scratches, cracks, and rust appear on complex curved surfaces, such as aircraft engine blades [9], bearing surfaces [13], ceramic surfaces [14], and thermos cup surfaces, they may shorten the product's lifespan or even lead to failure, posing safety hazards and the risk of accidents.

Many researchers have conducted studies on optical imaging and detection algorithms for complex curved surfaces. Zhou et al. [15] Employed dual strip backlights to illuminate both sides of highly reflective chrome-plated parts and fused the two images to reconstruct the surface profile. However, this separate illumination and fusion method is only effective for defects with certain surface undulations, such as scratches. When the object surface has surface contamination present or the defect undulations are subtle, the detection performance deteriorates. Xu et al. [16] proposed a defect detection method based on polarized light filtering, where they mathematically modeled the light propagation process and intensity loss. They suppressed image noise and enhanced defect contrast by leveraging polarization filtering. However, their experiments were conducted on simple cylindrical smooth surfaces, lacking research on imaging complex surfaces with varying curvatures. Wei et al. [17] utilized a digital micromirror device (DMD) to implement a pixel-level spatiotemporal modulation method for adaptive illumination in vision imaging, effectively suppressing halo interference and strong specular reflections on the object's surface. However, the use of DMD for pixel-level spatiotemporal modulation increases system complexity, requiring precise light source control and real-time feedback. This high complexity may lead to increased costs, making large-scale industrial applications less feasible. Meng et al. [18] employed Gray codes and a four-step phase-shifting method to determine the absolute phase of reflected images. They used an absolute phase gradient transformation method and affine transformation for posture correction, while template matching was applied for detecting diffuse reflection surface defects. Additionally, grayscale morphological opening and closing operations were performed on the original images to obtain defect shape and position information, thereby identifying image defects. However, on complex surfaces or at defect edge regions, reflected light may undergo multiple reflections or occlusions, leading to irregular phase variations, which in turn affect the accuracy of defect detection. Chen et al. [19] proposed a novel method for the imaging and edge detection of internal delamination defects in carbon fiber reinforced polymers (CFRP) using continuous line laser scanning excitation. Their Flash Laser Scanning Thermography (FLST) system captured thermal images and reconstructed the infrared images of defects using simple feature lines, while the effects of different parameters on the reconstructed images were analyzed. Finally, the morphological characterization of internal delamination defects in CFRP was achieved through a motion blur removal method combined with an edge detection algorithm. The results indicated that this method demonstrated high accuracy and efficiency in the non-destructive testing and evaluation of delamination defects.

High-precision optical imaging systems have laid a solid foundation for industrial inspection. However, in complex curved surface scenarios, accurate defect extraction on complex curved surfaces still faces two major challenges. First, defects are randomly distributed across curved surfaces, and their shapes can vary significantly depending on their location due to the varying geometry of the curved surface. This morphological diversity greatly increases the difficulty of defect extraction. Second, defects appear at multiple scales. Some defects are physically much smaller than the background region, resulting in a low signal-to-background ratio in the image, which raises the likelihood of missed detections. In contrast, larger defects with weak semantic features may be misclassified as background elements during feature extraction, also leading to detection errors. Currently, there are numerous defect extraction methods based on traditional vision algorithms. For instance, Bao et al. [20] designed a machine vision-based defect detection system for the surface of cylindrical rollers in bearings. By applying Otsu threshold segmentation, Canny edge detection, and morphological processing, they extracted and identified surface defects on rollers, classifying and labeling them accordingly. Wang et al. [21] proposed a dual-channel feature fusion-based vision detection method for the rapid detection of wheel hub cracks. Zheng et al. [22] introduced a method based on multi-wavelet transform and autoencoder networks for defect matching on steel surfaces. Liu et al. [23] developed a hybrid recognition method combining mathematical morphology and pattern recognition for defect detection on PCB surfaces.

In parallel, a wide range of defect detection methods based on deep learning have emerged, driving significant progress in object detection networks. These include single-stage object detection networks such as the YOLO series [24] and two-stage detection networks like Faster R-CNN [25]. However, deep learning-based defect detection often faces challenges such as sample imbalance and the difficulty of detecting small targets [26]. To address the challenge of small object detection in defect detection, subsequent researchers have integrated attention mechanisms and multi-scale feature fusion into neural networks to enhance their ability to detect small defects. Song et al. [27] proposed an cross-layer semantic guided network (CSGNet) based on YOLOv6 to detecting defects in small aero-engine blades. Zhao et al. [28] designed a defect detection model for turbine blades based on ShuffleNetV2 and Coordinate Attention. Li et al. [29] proposed a deep convolutional neural network (DCNN) for detecting surface defects on aero-engine blades. Liu et al. [30] introduced Bearing-DETR, an optimized deep learning model based on the RT-DETR architecture for bearing defect detection. Fei et al. [31] proposed TinyDefectNet, a specialized deep ensemble learning algorithm for the online identification of micro-defects in large images acquired from dedicated image acquisition modules.

To tackle the technical challenges associated with defect detection on complex curved surfaces, this study proposes several innovative solutions. First, the imaging of such surfaces is highly prone to specular glare, which can obscure critical defect information. To overcome this, polarized imaging technology is employed to suppress glare interference, thereby improving the signal-to-noise ratio of defect features. Second, to mitigate defect shape distortion induced by the geometric complexity of curved surfaces, a multi-scale edge information selection architecture is introduced to enhance the model's adaptability to morphologically diverse defects. Furthermore, a spatial pyramid pooling module is integrated to capture cross-scale contextual information, enriching the semantic representation and improving detection robustness.

2. Method

To address the challenges of glare interference and morphological diversity in defect detection on complex curved surfaces, this study proposes a dual-level solution that integrates polarization imaging with deep learning optimization. At the optical imaging level, polarization imaging technology is utilized, where polarizers are employed to filter out glare predominantly composed of polarized light, while retaining randomly scattered diffuse reflections that accurately convey the surface characteristics of curved objects. At the algorithmic level, a multi-scale edge information selection (MSIS) module and a Focal

Modulation module are embedded into the YOLOv11 architecture. By leveraging a parallel-branch network to extract edge features across multiple scales, coupled with an adaptive weighting mechanism for effective multi-scale feature fusion, the proposed method significantly improves the model's capacity to detect defects of varying morphologies on complex surfaces. Finally, a specialized dataset is constructed using polarization-filtered images with annotated defect labels, which is then used to train the enhanced MF-YOLOv11 model.

2.1. Polarization Imaging Principle

Uneven illumination on curved surfaces leads to variations in light reflection angles during imaging, resulting in inconsistent pixel intensities across different surface regions and increasing the difficulty of defect extraction. Moreover, reflective interference on smooth surfaces often generates glare, which can weaken or even obscure defect information in affected areas. Compared to conventional imaging techniques, polarization imaging offers distinct advantages in capturing material properties, surface roughness, and geometric features. This capability enhances the extraction of defect textures and structural details, thereby improving detection accuracy and robustness. The polarization state of light can be represented by the Stokes vector (I, Q, U, V), as defined in Equation (1). Here, I represents the total light intensity, Q denotes the intensity difference between horizontally and vertically polarized light, U represents the intensity difference between +45° polarized light and −45° polarized light, and V denotes the intensity difference between left- and right-handed circularly polarized light.

$$\begin{cases} I = I_t \\ Q = (I_{0^\circ} - I_{90^\circ}) \\ U = (I_{+45^\circ} - I_{-45^\circ}) \\ V = (I_{rc} - I_{lc}) \end{cases} \tag{1}$$

In this case, I_t represents the total light intensity, while I_0, I_{45}, I_{90}, and I_{-45} represent the intensity of polarized light at angles of 0°, 45°, 90°, and −45° to the horizontal direction, respectively. I_{rc} and I_{lc} represent the intensities of right circularly polarized and left circularly polarized light, respectively. The degree of linear polarization of light is defined by Equation (2) as follows:

$$y_{DoLP} = \frac{\sqrt{Q^2 + U^2}}{I_t} \tag{2}$$

When the light is completely composed of polarized light, the degree of linear polarization, calculated using Equation (2), is equal to one. On the other hand, the Stokes vector for natural light is represented as $(I_0,0,0,0)$, with a linear polarization degree of zero. By adding a linear polarizer to the surface of the light source, the natural light emitted by the source can be converted into linearly polarized light. According to Malus' law, the intensity of light after passing through the polarizer can be expressed by Equation (3):

$$I = \frac{1}{2\pi}\int_0^{2\pi} I_\theta \cos^2\theta d\theta = \frac{I_0}{2} \tag{3}$$

From Equation (3), it can be observed that natural light, which has a completely random vibration direction, will have its intensity reduced to half of I_0 after passing through a polarizer, where I_0 is the total light intensity. At this point, the light is fully converted into linearly polarized light. According to electromagnetic theory and Fresnel's law, when a light wave strikes the surface of a medium, its electric field vector can be decomposed into two orthogonal components: the s-component and the p-component. The s-component's electric field direction is perpendicular to the incident plane, while the p-component's electric field direction is parallel to the incident plane. The reflection coefficient r_s and transmission coefficient t_s for s-polarized light, as well as the reflection coefficient r_p and transmission coefficient t_p for p-polarized light, are defined in Equation (4).

$$\begin{cases} r_s = \frac{n_1 \cos\theta_i - n_2 \cos\theta_t}{n_1 \cos\theta_i + n_2 \cos\theta_t} \\ t_s = \frac{2n_1 \cos\theta_i}{n_1 \cos\theta_i + n_2 \cos\theta_t} \\ r_p = \frac{n_2 \cos\theta_i - n_1 \cos\theta_t}{n_2 \cos\theta_i + n_1 \cos\theta_t} \\ t_p = \frac{2n_1 \cos\theta_i}{n_2 \cos\theta_i + n_1 \cos\theta_t} \end{cases} \quad (4)$$

In Equation (4), n_1 denotes the refractive index of the incident medium, typically air, with a value of 1. n_2 represents the refractive index of the transmitted medium, set to 1.35 based on the common optical properties of curved materials. θ_i is the incident angle, and due to the light source's alignment perpendicular to the surface, the incident angle is 90 degrees. θ_t is the refraction angle, which is 47.8 degrees. According to Equation (4), both polarization components exhibit 100% reflectance, and the transmittance is 0%. As the incident light is completely linearly polarized, the reflected light is also fully linearly polarized. However, because the light source is a planar source, it not only illuminates the surfaces perpendicular to itself but also those with varying orientations. In these cases, the incident angle is not 90 degrees, and thus the reflectance is less than 100%, and the transmittance is greater than 0%. Consequently, non-linearly polarized components, such as diffusely reflected or elliptically polarized light, appear on the object's surface. These components retain random polarization states. The principle of the polarization imaging system is illustrated in Figure 1.

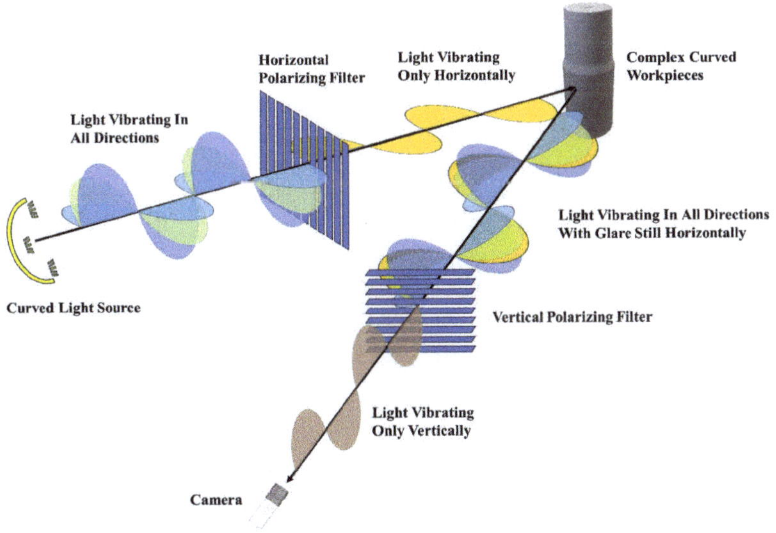

Figure 1. Polarization imaging system principle.

To suppress glare, primarily composed of linearly polarized light, while preserving useful surface information, a polarizer is employed to selectively block the linearly polarized component of reflected light and allow randomly or non-linearly polarized light to pass through. This effectively retains valuable image features. The underlying principle is illustrated in Figure 1. Given that the polarization direction of incident light is fixed, the linearly polarized portion of the reflected light tends to be more concentrated. By adjusting the rotation angle of the polarizer in front of the camera, the glare can be maximally attenuated, thereby enhancing the visibility of surface details on the object.

Therefore, a polarizer is first used to polarize the light emitted from the light source. Taking a horizontal polarizer as an example, the Stokes vector of the horizontally polarized light after polarization is represented as (I_0, I_0, 0, 0), indicating that the light remains horizontally polarized after reflecting off the object's surface. In this stage, a vertical polarizer is placed in front of the imaging device. Since the vertical polarizer is orthogonal to the polarization direction of the incident light, the intensity of the transmitted light is significantly reduced. The transmitted intensity can be described by Equation (5):

$$I = \frac{1}{2}(I_0 + Q_0 \cos^2(\theta) + U_0 \sin^2(\theta)) \tag{5}$$

From Equation (5), it can be observed that since the angle θ between the horizontally polarized light and the vertical polarizer is 90 degrees, the glare caused by the horizontally polarized component is effectively filtered out. As a result, only a portion of the incident light—primarily consisting of diffusely reflected and elliptically polarized light—is transmitted. This remaining light effectively captures and preserves the surface characteristics of the complex curved object.

2.2. *Improving the YOLOv11 Model*

This study proposes the MF-YOLOv11 detection model to address the challenges of diverse defect morphologies, small defect scales, and indistinct defect features in defect detection for complex curved surfaces. The proposed model incorporates the following innovative modules: (1) A C3k2-MSIS module based on multi-scale edge information selection (MSIS) is used, which processes features through adaptive pooling at different scales and depthwise separable convolutional branches combined with an edge enhancement module to refine edge details. The model's robustness is significantly improved by employing a dual-domain selection mechanism that balances spatial key region focusing and dynamic channel feature optimization. (2) The Focal Modulation module, which replaces the traditional SPPF module, is also used. This module leverages contextual relevance and a gating mechanism to dynamically assign weights, selectively aggregating key contextual information. This approach mitigates the indiscriminate fusion in conventional SPPF structures that often introduces noise interference, thereby enhancing the effectiveness of feature integration. Additionally, convolution kernels of different sizes are utilized to extract multi-scale features ranging from local to global levels, replacing traditional pooling operations to improve edge and detail capture capabilities. The improved MF-YOLOv11 structure is shown in Figure 2.

2.2.1. Multi-Scale Edge Information Selection Module

To enhance the network's ability to extract multi-scale features, a multi-scale edge information selection (MSIS) module is introduced based on the original C3k2 module. The module framework is shown in Figure 3. The core principle of this module is to perform adaptive pooling at different scales on the input image, followed by depthwise separable convolution branches to process features at each scale. After feature upsampling, an edge enhancement module is applied to amplify the edge information in the feature map. The resulting feature maps are concatenated along the channel dimension and then passed through a dual-domain selection mechanism module to focus on key features in both the spatial and channel dimensions.

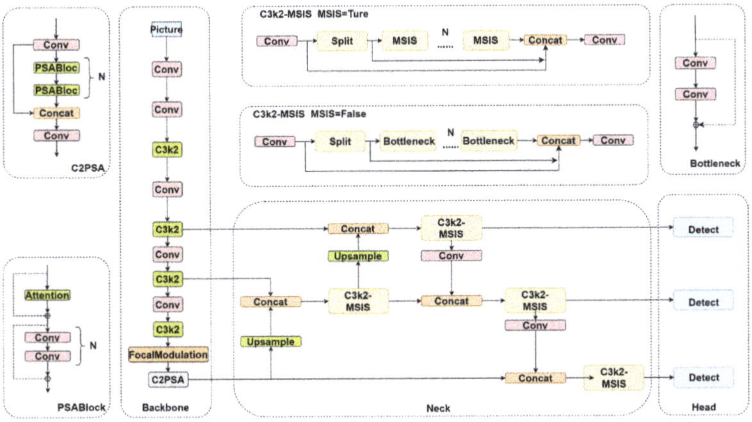

Figure 2. MF-YOLOv11 network architecture.

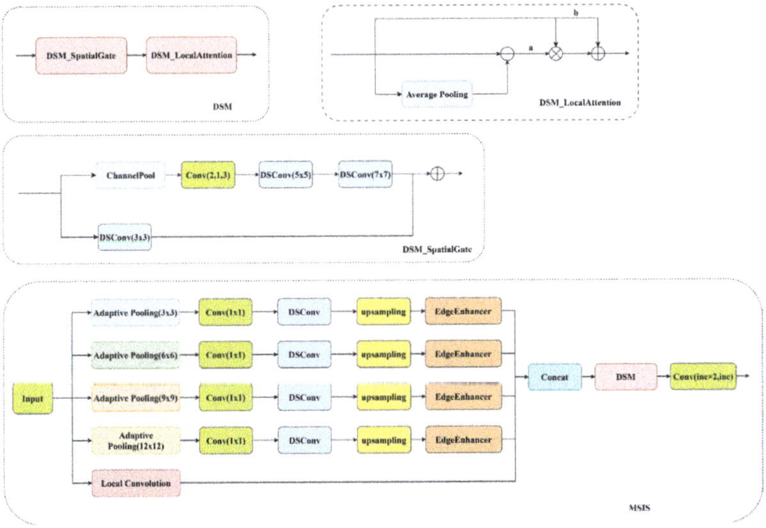

Figure 3. Multi-scale edge information selection module framework. a and b are learnable weights designed to enable adaptive adjustment of channel features. In the diagram, a indicates that the feature is multiplied by weight a during this process, while b denotes multiplication by weight b.

The edge enhancement module extracts the low-frequency information of the image through average pooling and subtracts this low-frequency information from the original input image x, thereby retaining the high-frequency information which corresponds to the edges. Simultaneously, the edge weight w is computed using a Sigmoid function. The edge information is then multiplied by the weight w and added back to the original input image x, enabling the preservation of the original information while emphasizing the edge details. The mathematical definitions of the overall process are given by Equations (6) and (7).

$$output = x + w(x - AvgPool(x)) \tag{6}$$

$$w = \frac{1}{1 + e^{-(x - AvgPool(x))}} \tag{7}$$

The dual-domain selection mechanism (DSM) consists of the spatial-domain selection module and the channel-domain selection module. The spatial-domain selection module utilizes a stacked structure of 5×5 and 7×7 convolutional kernels to capture structural information from the image, along with a 3×3 convolutional kernel to capture local detail information. Subsequently, global average pooling and global maximum pooling are performed in parallel on the input feature map, and their outputs are concatenated along the channel dimension to form a 2-channel feature that contains global statistical information. A subsequent 1×1 convolution layer is used to aggregate the 2 channels and generate a spatial attention weight map. This weight map is then multiplied by the image's structural information and added to the original local detail information through a residual connection.

The channel threshold selection module begins by initializing learnable weights a and b, which enable the adaptive adjustment of the channel features. The mean value of the spatial dimensions of the input feature x is then calculated to obtain a reference value that represents the global features. Subsequently, the input feature x is subtracted from this mean value to obtain the feature deviation out, establishing a weighted residual connection structure. The specific definition of this process is provided in Equation (8).

$$output = a \times out \times x + b \times x \tag{8}$$

The learnable weights a and b dynamically adjust the fusion ratio between the original features and the feature deviation, enhancing the model's ability to represent feature distribution shifts while preserving the original feature information.

2.2.2. Focal Modulation Module

The core principle of the Focal Modulation module is hierarchical context extraction and dynamic gating mechanism. Its function is to replace the self-attention mechanism by utilizing depthwise separable convolutions to capture multi-scale visual context information from local to global scale. The gating mechanism then dynamically aggregates key contextual information, thereby reducing computational complexity while enhancing the model's ability to perceive multi-scale features. The dynamic gating mechanism achieves the adaptive fusion of multi-scale features through dynamic weights g_l. The model principle of dynamic gating is illustrated in Figure 4.

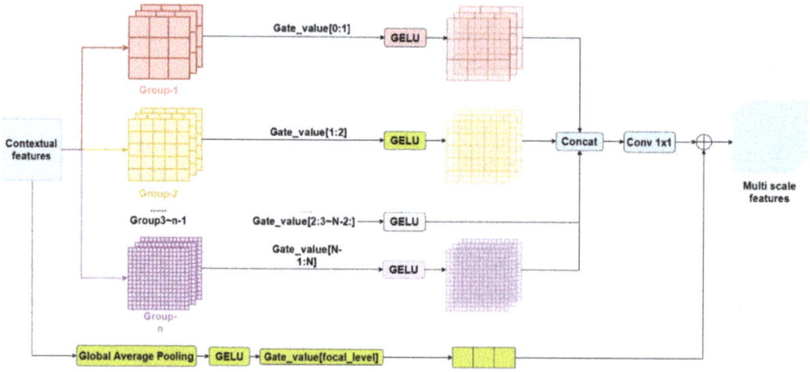

Figure 4. Gating mechanism principle.

Each group consists of convolutional kernels of different sizes, which are used to extract feature information at different scales. The extracted feature information is multiplied by the corresponding dynamic weight g_l and passed through a GELU activation function to introduce non-linearity. The approximate definition of the GELU activation function is given in Equation (9). The activated feature maps are then concatenated along the channel dimension, and multi-scale feature fusion is completed using a 1 × 1 convolution.

$$\mathrm{GELU}(x) = 0.5x\left[1 + \tanh\left(\sqrt{\frac{2}{\pi}}(x + 0.047715x^3)\right)\right] \tag{9}$$

The green process box below Figure 4 indicates that the global average pooling is used to obtain the spatial mean of the contextual information, which is then added to the feature map after multi-channel feature fusion, resulting in the aggregated contextual information ctx_{all}. The expression is given in Equation (10).

$$ctx_{all} = \sum_{l=0}^{L-1} Conv_l(ctx)g_l + GAP(ctx)g_L \tag{10}$$

The Focal Modulation module demonstrates significant advantages over the traditional SPPF structure. Firstly, it replaces fixed-size pooling layers with hierarchical convolution, enabling more flexible multi-scale edge and detail feature extraction. Secondly, its gated aggregation mechanism emphasizes content relevance, effectively mitigating the noise interference caused by indiscriminate feature fusion in SPPF. Furthermore, Focal Modulation reduces the computational complexity from $O(N^2)$ in SPPF to a linear complexity of O(N), maintaining the capability of multi-scale feature fusion while significantly improving computational efficiency, which is particularly beneficial for high-resolution image processing tasks.

3. Experimental Results and Analysis

3.1. Test Equipment Construction

To verify the effectiveness of the proposed polarization imaging method and the enhanced MF-YOLOv11 model, a series of complex curved surface samples were fabricated for testing, and a dedicated inspection platform was constructed. The hardware configuration of the inspection platform is shown in Figure 5. It includes a rotational and fixation mechanism as well as multiple imaging modules. The detailed hardware specifications of the image acquisition unit are listed in Table 1.

Table 1. Image acquisition module hardware composition.

Number	Device Name	Brand	Model
1	Camera	Basler, Ahrensburg, Germany	Aca-2500-30gc
2	Lenses	ZLZK, Zoomlion, Changsha, China	LM0820MP5
3	Light Source	V-Light, China	VLHBGLXD30X245B-24V
4	Industrial Computer	Hikvision, Hangzhou, China	MV-IPC-F487H
5	PLC	Siemens, Munich, Germany	S7-200 SMART
6	Frame grabber	Daheng Imaging, Beijing, China	PCIe-GIE72P

The fixation module, as illustrated in Figure 5a, is fitted with elastic friction pads on its outer surface. When a sample is placed and secured, the pads deform to provide both clamping force and friction, ensuring the stable fixation of complex curved surfaces. The servo motor, shown in Figure 5b, is installed beneath the inspection platform and connected to a reducer. The output shaft of the reducer is coupled with the fixation module, enabling the sample to rotate during a full circumferential scan. At the center of the inspection platform, a 400 mm × 600 mm arc-shaped light source is installed. Its light-emitting surface is covered by a polarizer of the same dimensions. Two through-holes are located at the center of the arc light source, allowing the polarization camera to capture images of the sample through them. As shown in Figure 5c, the polarization camera is mounted behind the arc light source on a supporting bracket. In addition, two side cameras are positioned on the left and right sides of the platform, as illustrated in Figure 5d. These cameras are used to detect non-planar defects on the sample surface. To ensure sufficient and uniform illumination, a ring light and two strip lights—mounted on either side of the ring—are employed, as shown in Figure 5e.

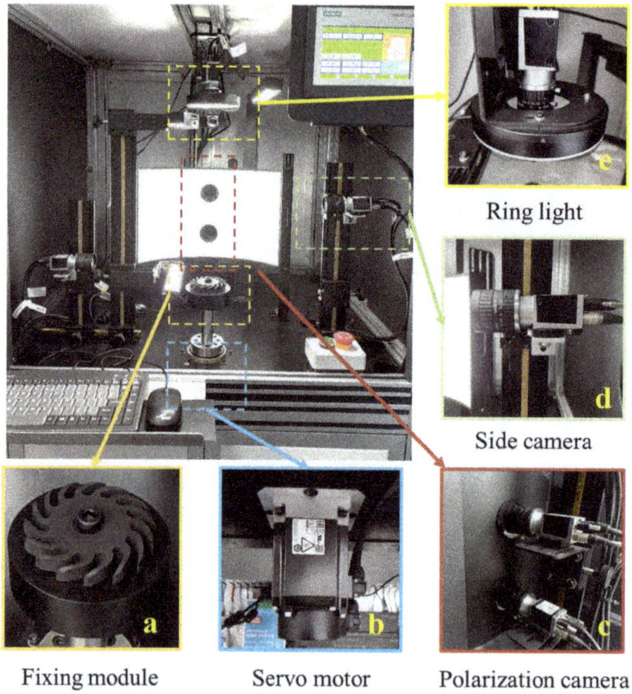

Figure 5. Photographs of experimental equipment. (**a**) Fixing module. (**b**) Servo motor. (**c**) Polarization camera. (**d**) Side camera. (**e**) Ring light.

To obtain a complete image sequence of the sample, the servo motor rotates in 90° increments, briefly pausing in each step. During each pause, all imaging units simultaneously capture images of the sample. Once imaging is completed in the current position, the servo motor advances by another 90°, and the process repeats. This cycle continues until a full 360° image sequence of the sample is acquired.

3.2. Polarization Imaging Performance Experiment

To evaluate the effectiveness of polarization imaging, 100 raw images without polarization filtering and 100 images captured with polarization imaging for glare suppression were acquired using the equipment described in Section 3.1, under consistent lighting and camera settings. These images were compared against results obtained using the high dynamic range imaging (HDRI) algorithm, a commonly applied technique for defect detection on highly reflective metal surfaces. In the context of defect detection, image quality is primarily assessed using three key metrics: root mean square (RMS) noise, standard deviation, and the uniformity index.

The definition of root mean square (RMS) noise is given in Equation (11), where $F(i,j)$ represents the test image and $F'(i,j)$ represents the reference image, which is obtained by applying Gaussian low-pass filtering to the test image. The definition of Gaussian low-pass filtering is provided in Equation (12). Low-pass filtering effectively removes noise from the image. RMS noise reflects the intensity of the high-frequency noise removed; a higher value indicates a greater presence of high-frequency noise components in the original image, and consequently, a higher amount of noise energy removed after filtering.

$$\text{RMS Noise} = \sqrt{\frac{1}{MN}\sum_{i=1}^{M}\sum_{j=1}^{N}(F(i,j) - F'(i,j))^2} \tag{11}$$

$$G(x,y) = \frac{1}{2\pi\sigma^2}e^{-\frac{x^2+y^2}{2\sigma^2}} \tag{12}$$

The definitions of image standard deviation and uniformity index are given in Equations (13) and (14), respectively. In Equation (13), u represents the mean grayscale value of the image. The standard deviation measures the degree of dispersion of pixel values relative to the mean grayscale value, reflecting the concentration or spread of pixel distributions. A lower standard deviation indicates a more concentrated pixel distribution, suggesting the effective removal of high-intensity glare interference. Conversely, a higher standard deviation indicates a more dispersed pixel distribution, implying the poor suppression of high-intensity glare interference. The uniformity index is calculated as the sum of the squared probabilities of each grayscale level and directly reflects the concentration of the grayscale distribution. A higher uniformity index indicates that the grayscale values are concentrated within a few levels, implying strong grayscale consistency in the captured image, which is beneficial for subsequent algorithmic processing. In contrast, a lower uniformity index suggests that grayscale values are more scattered, indicating the presence of significant glare interference, which is detrimental to further image analysis.

$$\text{Standard Deviation} = \sqrt{\frac{1}{MN}\sum_{i=1}^{M}\sum_{j=1}^{N}(F(i,j) - u)^2} \tag{13}$$

$$\text{Uniformity Index} = \sum_{i=0}^{255}(p_i)^2 \tag{14}$$

Figure 6 presents the mean values of key metrics for the original image, the image processed by the HDRI algorithm, and the image obtained using polarization imaging. As shown in Figure 6a, the polarization imaging method achieved a lower root mean square (RMS) noise value than both the original and HDRI-processed images. Specifically, the RMS noise of the polarization image was 1.85 lower than that of the original image and 0.96 lower than that of the HDRI-processed image. This demonstrates that polarization imaging effectively reduced image noise, primarily by significantly mitigating glare interference. Furthermore, polarization imaging resulted in a lower standard deviation compared to both the original and HDRI-processed images. Standard deviation serves as an indirect measure of image uniformity, with lower values indicating higher uniformity. As illustrated in Figure 6b, the standard deviation of the polarization image was 20.57 lower than that of the original image and 9.73 lower than that of the HDRI-processed image, underscoring its enhanced uniformity. To further evaluate image uniformity, the uniformity index was employed to quantify grayscale consistency. As shown in Figure 6c, the uniformity index of the polarization image increased by 0.31 compared to the original image and by 0.16 compared to the HDRI-processed image, indicating a substantial improvement in grayscale uniformity.

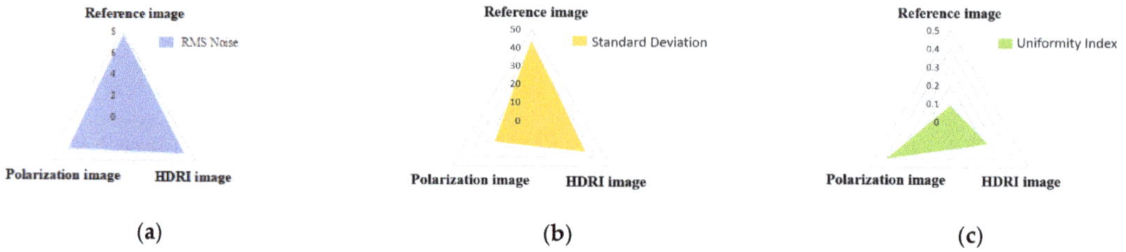

Figure 6. Image quality evaluation indicators. (**a**) RMS noise; (**b**) standard deviation; (**c**) uniformity index.

In summary, the proposed polarization imaging method effectively enhances image uniformity and reduces noise, leading to improved performance in defect detection algorithms. Furthermore, when compared to the HDRI method, which requires capturing multiple sets of images at varying exposure levels, polarization imaging significantly reduces the time needed for image acquisition, thereby boosting the overall efficiency of defect detection.

3.3. *MF-YOLOv11 Model Performance Experiment*

To validate the reliability of the improved YOLOv11 model in detecting defects on complex curved surfaces, a series of samples with varying curvatures were fabricated, and surface defect images were systematically collected. The dataset encompasses a variety of defect types and sizes, including pits, discoloration, and surface particles. Prior to image acquisition, a polarization-based imaging technique was employed to suppress glare on reflective surfaces, ensuring higher image quality. A total of 1536 defect images were captured, from which 1228 non-redundant samples were selected for training, and the remaining 308 images were used for validation. To improve the model's generalization capability and robustness in handling diverse defect scenarios, data augmentation techniques were applied to the training set. These included geometric transformations, color space adjustments, and noise injection, ultimately expanding the training dataset to 4912 images. The hardware setup and development environment used in the experiments are summarized in Table 2.

Table 2. Model parameter configuration.

Name	Parameter
CPU	Intel i9-13900HX
GPU	NVIDIA RTX 4060
WINDOW	WINDOW11
PYTHON	3.9
CUDA	11.7
TORCH	2.0.1
Learning rates	0.01
Epochs	250
Batch size	16

3.3.1. MF-YOLOv11 Model Checking Performance Evaluation

To evaluate the performance of the model, the following evaluation metrics were used in this experiment: accuracy, which reflects the proportion of actual positive samples among those predicted as positive; recall, which assesses the coverage of correctly identified true positive samples; mAP50, the mean average precision calculated at an IoU threshold of 0.5; and mAP50-95, the mean average precision calculated across IoU thresholds from 0.5 to 0.95, which comprehensively reflects the localization and classification ability for multi-class defect detection. The definitions of these evaluation metrics are given in Formulas (12) and (13). In the formula definition, *TP* stands for true positive, *FP* stands for false positive, *FN* stands for false negative, and *TN* stands for true negative.

$$Precision = \frac{TP}{TP + FP} \tag{15}$$

$$Recall = \frac{TP}{TP + FN} \tag{16}$$

The training of the MF-YOLOv11 model was conducted using the PyTorch v2.0.1 framework, with a batch size of 16 and a learning rate of 0.01, over a total of 250 epochs. The results of the training process are shown in Figure 7. As illustrated in Figure 7a, precision significantly improved after the 20th epoch and stabilized around the 150th epoch. Recall, presented in Figure 7b, also exhibited a notable increase after 20 epochs, continuing to rise steadily and stabilizing around the 150th epoch. The mAP50 and mAP50-95 metrics increased rapidly during the initial stages of training, followed by a noticeable dip around the 130th epoch, after which they continued to rise steadily and stabilized after approximately 200 epochs. These observations suggest that both precision and recall improved significantly early in the training process, indicating that the model quickly learned effective defect detection features, stabilizing after 150 epochs. As the training progressed, all metrics reached stability after around 200 epochs, indicating that the model had fully converged and that the training parameters were conducive to achieving consistent performance.

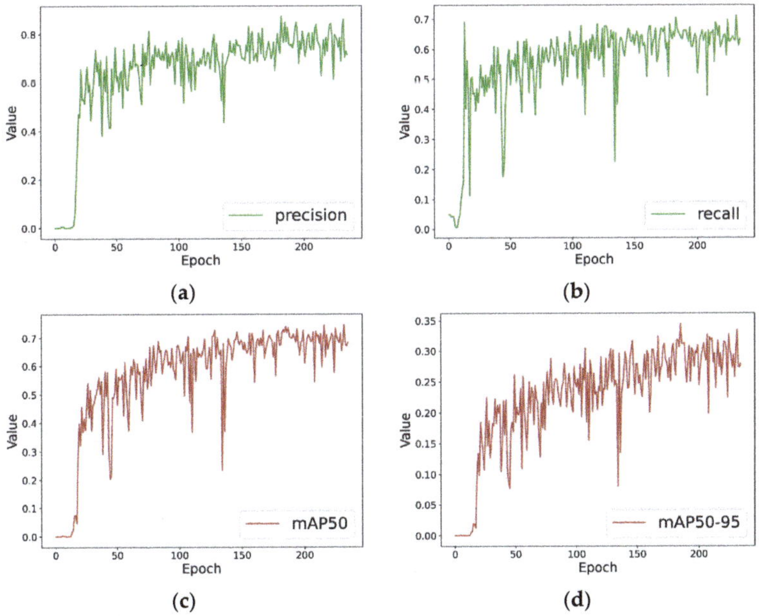

Figure 7. Training evaluation metrics. (**a**) Precision; (**b**) recall; (**c**) mAP50; (**d**) mAP50-95.

In addition to the model performance metrics mentioned above, the loss function used for optimizing the model is also of significant importance, as it reflects the degree of difference between the predicted results and the ground truth. The loss function includes bounding box regression loss, classification loss, and dynamic feature loss, with the specific results shown in Figure 8.

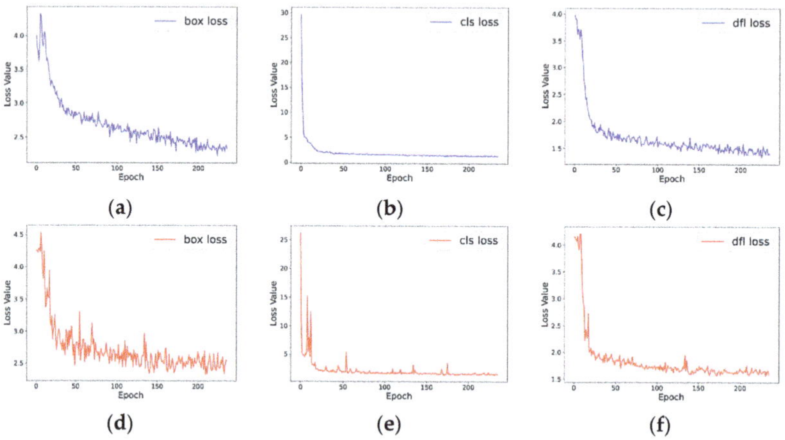

Figure 8. Model loss function. (**a**) Training set box loss; (**b**) training set classification loss; (**c**) training set distribution focal loss (DFL); (**d**) validation set box loss; (**e**) validation set classification loss; (**f**) validation set distribution focal loss (DFL).

Figure 8a–c illustrate the training loss curves, reflecting the model's fitting performance on the training data. Figure 8d–f depict the validation loss curves, assessing the model's generalization ability on unseen data. Figure 8a shows the bounding box regression loss curve during training. After a brief increase in the early stages, the loss began to decrease and stabilized at around 200 epochs. Figure 8b illustrates the classification loss during training, which stabilized after 20 epochs and remained steady thereafter. Figure 8c depicts the dynamic feature loss during training, which decreased rapidly in the early stages, then slowed down after 30 epochs, and stabilized after 200 epochs. Figure 8d presents the bounding box regression loss on the validation set, which decreased rapidly in the early stages, then declined more gradually after 25 epochs, and stabilized after 150 epochs. The trend in the validation set aligns with that of the training set, indicating that the model's generalization ability for localization tasks improved progressively and converged. Figure 8e illustrates the classification loss on the validation set, which stabilized after 20 epochs, with minor fluctuations afterward. Figure 8f depicts the dynamic feature loss on the validation set, which declined rapidly in the early stages and stabilized after 200 epochs, mirroring the trend observed in the training set. This indicates that the model's performance on the validation set followed a similar trend to that of the training set, suggesting robust feature extraction capabilities on unseen data, with no significant indications of overfitting.

The optimal model achieved an accuracy of 86.1%, a recall rate of 71.1%, and an mAP50 of 72.7%. To visually assess the model's detection performance, a subset of images was randomly selected from a validation set of 308 images, and the trained optimal model was applied for detection. The images in Figure 9a–c were captured without using polarization imaging. The yellow lines in these images highlight the curvature of the samples, simulating changes in complex surfaces by varying the curvature. Figure 9d–f display images obtained using polarization imaging, with defect detection results from the MF-YOLOv11 model. A visual comparison clearly demonstrates that polarization imaging effectively suppressed glare interference, preventing the loss of defect information and enhancing the overall uniformity of the imaging. It can be observed that the detection model accurately marked the defect locations, showcasing the model's strong detection capability.

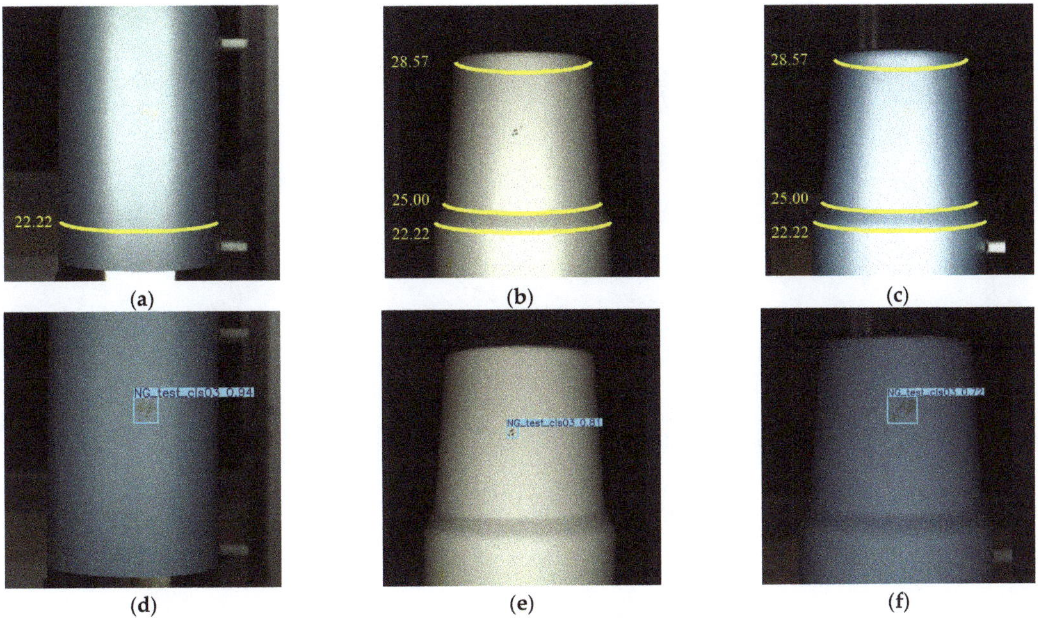

Figure 9. Detection effect of plane defects. (**a**) Sample 1 without polarization imaging; (**b**) Sample 2 without polarization imaging; (**c**) Sample 3 without polarization imaging; (**d**) Sample 1 with polarization imaging; (**e**) Sample 2 with polarization imaging; (**f**) Sample 3 with polarization imaging.

In addition to planar defects detected using polarization imaging, complex curved surfaces also contain numerous non-planar defects, which can be categorized into protruding defects and pit defects. Protruding defects are characterized by small defect scales, where the small targets occupy only a few pixels in the image, making it challenging for traditional convolutional neural networks to capture fine details. On the other hand, pit defects have larger defect scales but weaker semantic information, low contrast with the background, and more blurred boundaries. The proposed MF-YOLOv11 model, incorporating a multi-scale edge information selection module and an improved spatial pyramid pooling module, not only enhances the semantic information of defect edges but also adaptively fuses multi-scale features, improving the model's detection capability for both types of defects. The specific detection results are as follows: Figure 10a–c show the detection labels for protrusion defects using the MF-YOLOv11 model, while Figure 11a–c show the detection labels for pit defects. The model demonstrated strong detection capabilities for both types of non-planar defects.

3.3.2. MF-YOLOv11 Model Detection Performance Comparison

From Figure 12, it can be seen that the detection accuracy of the Faster R-CNN model was relatively low, with a value of only 70.7%. Additionally, the FLOPs per inference were 17.9, indicating a high computational cost for inference. The improved network models based on DETR exhibited better detection performance. The RT-DETR model achieved a detection accuracy of 76.7%, and the Bearing-DETR model achieved 78.7%. However, both models required substantial computational resources, with a single inference time of around 2 s, which is not suitable for real-time defect detection. The CSGNet model is an improved version of the YOLOv6 model, offering a good balance between detection accuracy and computational speed. The Net-CA-SSD model, which uses ShuffleNetv2 as the backbone for the SSD detection framework, achieved a detection accuracy of 76.8%. However, like the improved DETR-based networks, it also suffered from longer inference times.

(a) (b) (c)

Figure 10. Detection effect of protrusion defects. (**a**) Detection result of Sample 1; (**b**) Detection result of Sample 2; (**c**) Detection result of Sample 3.

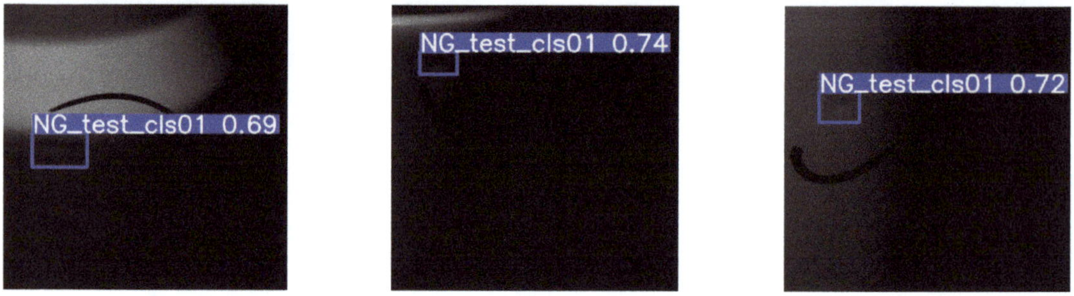

(a) (b) (c)

Figure 11. Detection effect of pit defects. (**a**) Detection result of Sample 1; (**b**) Detection result of Sample 2; (**c**) Detection result of Sample 3.

Compared to other models, the MF-YOLOv11n model demonstrates significant advantages, particularly in terms of both performance and efficiency. First, this model outperforms others in key detection metrics, with a precision of 86.1% and a recall of 71.1%, both of which are the highest values among all models. This indicates that it effectively reduces both false positives and false negatives. Its mAP50 score of 72.7% also surpasses that of other models, suggesting superior balance between bounding box localization accuracy and classification precision. This is crucial for high-precision detection requirements in industrial quality control. Secondly, the MF-YOLOv11n achieves a lightweight design while maintaining high performance. The model size is only 5.8 MB, significantly smaller than Faster-RCNN (27.4 MB) and RT-DETR (32.2 MB), and slightly smaller than the same series YOLOv11n (5.9 MB). Its FLOPs value is 7.6, which is relatively low, indicating that the model consumes fewer computational resources and is better suited for deployment on industrial devices or embedded systems with limited computational power. Compared to YOLOv10n and YOLOv11n, MF-YOLOv11n further enhances its ability to capture defect features by optimizing the network structure, making it better suited for adapting to the diverse defects present on complex surfaces. In conclusion, MF-YOLOv11 outperforms existing models in terms of higher detection accuracy, lower resource consumption, and stronger feasibility for industrial implementation.

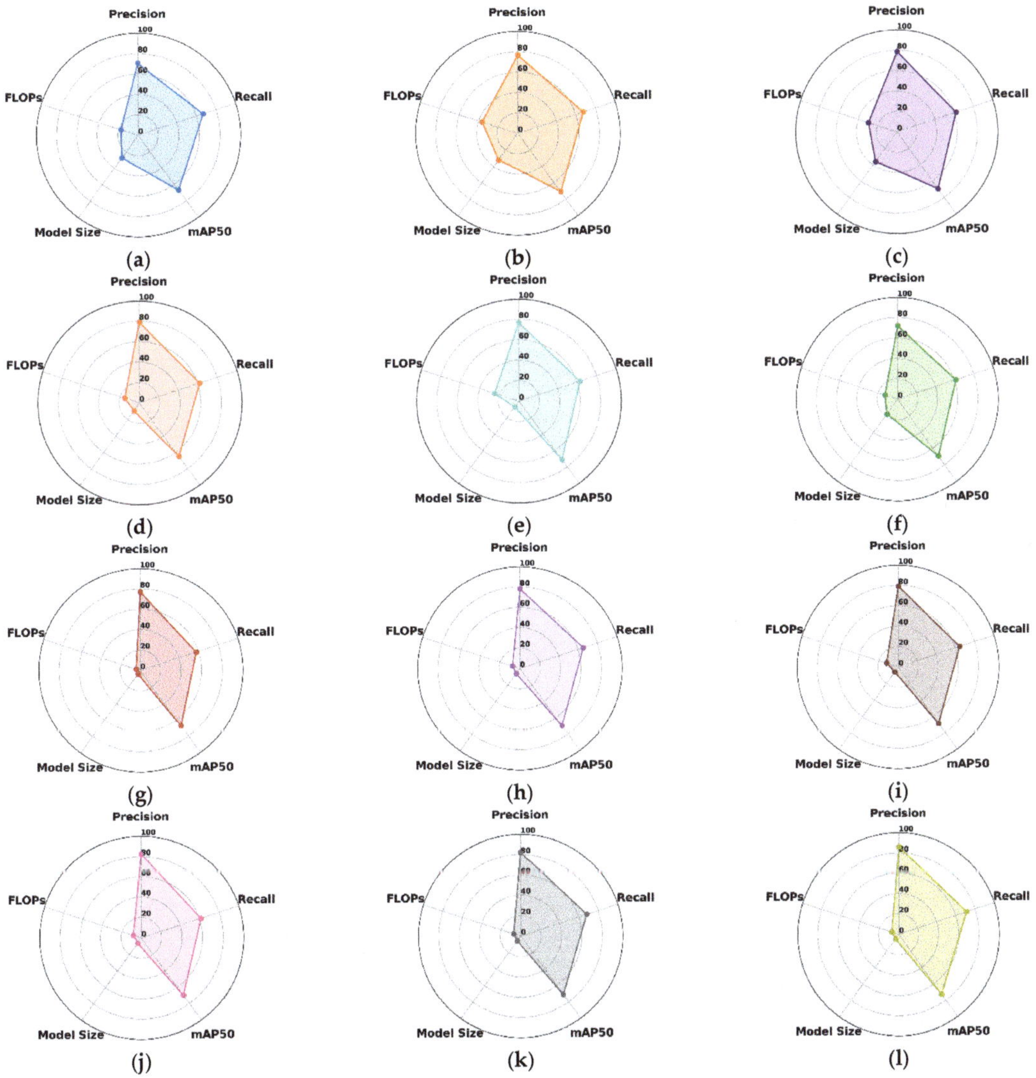

Figure 12. Model performance comparison. (**a**) Faster R-CNN; (**b**) RT-DETR; (**c**) Bearing-DETR; (**d**) CSGNet; (**e**) Net-CA-SSD; (**f**) YOLOV3n; (**g**) YOLOV5n; (**h**) YOLOV8n; (**i**) YOLOv9n; (**j**) YOLOV10n; (**k**) YOLOV11n; (**l**) MF-YOLOv11n.

3.3.3. Ablation Study

This study systematically conducted ablation experiments to verify the performance improvement of the MF-YOLOv11n model by incorporating the multi-scale edge information selection (MSIS) module and the Focal Modulation module. The experiments were performed on a self-constructed complex surface defect dataset, with key hyperparameters such as a fixed batch size of 16 and an initial learning rate of 0.01 to minimize the influence of extraneous variables. The baseline model used was the original YOLOv11n architecture. Comparative experiments were conducted by replacing the C3k2 module in the neck with the C3k2-MSIS module and substituting the SPPF module in the backbone with the Focal Modulation module. The YOLOv11n model with only the C3k2-MSIS module replacement was designated as YOLOv11n-C3k2-MSIS, while the model with only the Focal Modulation module replacement was named YOLOv11n-Focal Modulation. The experimental results are summarized in Table 3.

Table 3. Ablation experiment results.

Model Name	Precision	Recall	mAP50	Model Size	GFLOPS
YOLOv11n	82.2	68.3	71.1	5.9	7.3
YOLOv11n-C3k2-MSIS	83.7	69.2	71.6	5.8	7.4
YOLOv11n-Focal Modulation	84.3	69.7	71.8	5.8	7.4
MF-YOLOv11n	86.1	71.1	72.7	5.8	7.6

It can be observed that, compared to the YOLOv11n baseline model, the introduction of the C3k2-MSIS module improved the model's accuracy by 1.5%, reaching 83.7%, recall by 0.9%, reaching 69.2%, and mAP50 by 0.5%, reaching 71.6%. On the other hand, the model with the Focal Modulation module showed more significant improvements, with accuracy increasing by 2.1% to 84.3%, recall increasing by 1.4% to 69.7%, and mAP50 increasing by 0.7% to 71.8%. When both improvement strategies were integrated into MF-YOLOv11n, the model achieved the best overall performance while maintaining a model complexity similar to that of the baseline model. This indicates that the two modules complement each other in enhancing edge feature extraction and dynamically focusing on key areas, effectively balancing detection accuracy and computational efficiency.

4. Conclusions

This study addresses the challenge of defect detection on the surfaces of industrial components with complex curvatures by proposing a combined approach that integrates a polarization imaging system with an improved MF-YOLOv11 model. The polarization imaging system employs a curved light source specifically designed to adapt to complex surfaces, ensuring uniform illumination. Additionally, a polarization imaging unit is constructed, where horizontal and vertical polarizers work in tandem to capture images in a single exposure while effectively suppressing glare interference. The polarization imaging method outperforms other approaches across all evaluation metrics. The MF-YOLOv11 model incorporates a multi-scale edge information selection module and a Focal Modulation module. The multi-scale edge information selection module employs adaptive pooling and parallel convolutional branches to process features at different scales, utilizing an edge enhancement module to enhance edge information and achieve multi-scale feature fusion. The Focal Modulation module applies hierarchical convolution to extract multi-scale contextual information and integrates a dynamic gating mechanism for feature aggregation. These module enhancements significantly improve the model's ability to detect defects of various forms on complex surfaces. Experimental results demonstrate that, on the self-constructed dataset, the MF-YOLOv11 model achieved a precision of 86.1%, reflecting a 3.9% improvement over the baseline YOLOv11n model. The recall reached 71.1%, an increase of 2.8% over the baseline, while the mAP50 metric achieved 72.7%, improving by 1.6%. Furthermore, the inference time per sample was less than 35 ms, meeting the real-time requirements for industrial defect detection.

Author Contributions: Conceptualization, Z.Y.; Methodology, Z.Y. and H.W.; Validation, D.W.; Data curation, Z.Y.; Writing—original draft, Z.Y.; Writing—review & editing, D.W. and H.W.; Visualization, H.W.; Supervision, H.W.; Funding acquisition, D.W. All authors have read and agreed to the published version of the manuscript.

Funding: This research was supported by the Zhejiang Provincial Natural Science Foundation of China under Grant No. LTGC23F030001 and Jinhua City Science and Technology Program Initiative Project (2023-1-001a)—Industrial Proactive Design, Research and Development of Vision Inspection Equipment for Engine Camshaft Defects.

Data Availability Statement: The original contributions presented in this study are included in the article. Further inquiries can be directed to the corresponding author.

Conflicts of Interest: The authors declare no conflicts of interest.

References

1. Zarei, A.; Pilla, S. Laser ultrasonics for nondestructive testing of composite materials and structures: A review. *Ultrasonics* **2024**, *136*, 107163. [CrossRef] [PubMed]
2. Li, E.; Guo, W.; Cao, X.; Zhu, J. A magnetic head-based eddy current array for defect detection in ferromagnetic steels. *Sens. Actuators A Phys.* **2024**, *379*, 115862. [CrossRef]
3. Chang, H.-L.; Ren, H.-T.; Wang, G.; Yang, M.; Zhu, X.-Y. Infrared defect recognition technology for composite materials. *Front. Phys.* **2023**, *11*, 1203762. [CrossRef]
4. Ren, Z.; Fang, F.; Yan, N.; Wu, Y. State of the Art in Defect Detection Based on Machine Vision. *Int. J. Precis. Eng. Manuf.-Green Technol.* **2021**, *9*, 661–691. [CrossRef]
5. Wen, X.; Shan, J.; He, Y.; Song, K. Steel Surface Defect Recognition: A Survey. *Coatings* **2022**, *13*, 17. [CrossRef]
6. Kim, T.; Behdinan, K. Advances in machine learning and deep learning applications towards wafer map defect recognition and classification: A review. *J. Intell. Manuf.* **2022**, *34*, 3215–3247. [CrossRef]
7. Xie, Y.; Xu, X.; Liu, S. Machine vision-based detection of surface defects in cylindrical battery cases. *J. Energy Storage* **2024**, *101*, 113949. [CrossRef]
8. Qiao, J.; Sun, C.; Cheng, X.; Yang, J.; Chen, N. Stainless steel cylindrical pot outer surface defect detection method based on cascade neural network. *Meas. Sci. Technol.* **2023**, *35*, 036201. [CrossRef]
9. Abdulrahman, Y.; Mohammed Eltoum, M.A.; Ayyad, A.; Moyo, B.; Zweiri, Y. Aero-engine Blade Defect Detection: A Systematic Review of Deep Learning Models. *IEEE Access* **2023**, *11*, 53048–53061. [CrossRef]
10. Jing, J.; Liu, S.; Wang, G.; Zhang, W.; Sun, C. Recent advances on image edge detection: A comprehensive review. *Neurocomputing* **2022**, *503*, 259–271. [CrossRef]
11. Tsai, D.-M.; Huang, C.-K. Defect Detection in Electronic Surfaces Using Template-Based Fourier Image Reconstruction. *IEEE Trans. Compon. Packag. Manuf. Technol.* **2019**, *9*, 163–172. [CrossRef]
12. Li, F.; Yuan, L.; Zhang, K.; Li, W. A defect detection method for unpatterned fabric based on multidirectional binary patterns and the gray-level co-occurrence matrix. *Text. Res. J.* **2019**, *90*, 776–796. [CrossRef]
13. Sahu, D.; Dewangan, R.K.; Matharu, S.P.S. An Investigation of Fault Detection Techniques in Rolling Element Bearing. *J. Vib. Eng. Technol.* **2023**, *12*, 5585–5608. [CrossRef]
14. Dong, G.; Pan, X.; Liu, S.; Wu, N.; Kong, X.; Huang, P.; Wang, Z. A review of machine vision technology for defect detection in curved ceramic materials. *Nondestruct. Test. Eval.* **2024**, 1–27. [CrossRef]
15. Zhou, A.; Ai, B.; Qu, P.; Shao, W. Defect detection for highly reflective rotary surfaces: An overview. *Meas. Sci. Technol.* **2021**, *32*, 062001. [CrossRef]
16. Xu, L.M.; Yang, Z.Q.; Jiang, Z.H.; Chen, Y. Light source optimization for automatic visual inspection of piston surface defects. *Int. J. Adv. Manuf. Technol.* **2016**, *91*, 2245–2256. [CrossRef]
17. Shao, W.; Liu, K.; Shao, Y.; Zhou, A. Smooth Surface Visual Imaging Method for Eliminating High Reflection Disturbance. *Sensors* **2019**, *19*, 4953. [CrossRef]
18. Hui, M.; Li, M.; Babar, M. A Diffuse Reflection Approach for Detection of Surface Defect Using Machine Learning. *Mob. Inf. Syst.* **2022**, *2022*, 7771178. [CrossRef]
19. Chen, H.; Gao, J.; Zhang, Z.; Yin, W.; Dong, N.; Zhou, G.; Meng, Z. CFRP delamination defect detection by dynamic scanning thermography based on infrared feature reconstruction. *Opt. Lasers Eng.* **2025**, *187*, 108884. [CrossRef]
20. Bao, Y.; Zhou, Z.; Wei, S.; Xiang, J. Research on surface defect detection system and method of train bearing cylindrical roller based on surface scanning. *J. Mech. Sci. Technol.* **2023**, *37*, 4507–4519. [CrossRef]
21. Wang, D.; Yin, J.; Wu, H.; Ge, B. Method for detecting internal cracks in joints of composite metal materials based on dual-channel feature fusion. *Opt. Laser Technol.* **2023**, *162*, 109263. [CrossRef]
22. Zheng, X.; Liu, W.; Huang, Y. A Novel Feature Extraction Method Based on Legendre Multi-Wavelet Transform and Auto-Encoder for Steel Surface Defect Classification. *IEEE Access* **2024**, *12*, 5092–5102. [CrossRef]
23. Liu, Z.; Qu, B. Machine vision based online detection of PCB defect. *Microprocess. Microsyst.* **2021**, *82*, 103807. [CrossRef]

24. Hussain, M. YOLO-v1 to YOLO-v8, the Rise of YOLO and Its Complementary Nature toward Digital Manufacturing and Industrial Defect Detection. *Machines* **2023**, *11*, 677. [CrossRef]
25. Ren, S.; He, K.; Girshick, R.; Sun, J. Faster R-CNN: Towards Real-Time Object Detection with Region Proposal Networks. *IEEE Trans. Pattern Anal. Mach. Intell.* **2017**, *39*, 1137–1149. [CrossRef]
26. Ameri, R.; Hsu, C.-C.; Band, S.S. A systematic review of deep learning approaches for surface defect detection in industrial applications. *Eng. Appl. Artif. Intell.* **2024**, *130*, 107717. [CrossRef]
27. Song, K.; Sun, X.; Ma, S.; Yan, Y. Surface Defect Detection of Aeroengine Blades Based on Cross-Layer Semantic Guidance. *IEEE Trans. Instrum. Meas.* **2023**, *72*, 1–11. [CrossRef]
28. Zhao, H.; Gao, Y.; Deng, W. Defect Detection Using Shuffle Net-CA-SSD Lightweight Network for Turbine Blades in IoT. *IEEE Internet Things J.* **2024**, *11*, 32804–32812. [CrossRef]
29. Li, D.; Li, Y.; Xie, Q.; Wu, Y.; Yu, Z.; Wang, J. Tiny Defect Detection in High-Resolution Aero-Engine Blade Images via a Coarse-to-Fine Framework. *IEEE Trans. Instrum. Meas.* **2021**, *70*, 1–12. [CrossRef]
30. Liu, M.; Wang, H.; Du, L.; Ji, F.; Zhang, M. Bearing-DETR: A Lightweight Deep Learning Model for Bearing Defect Detection Based on RT-DETR. *Sensors* **2024**, *24*, 4262. [CrossRef]
31. Chang, F.; Liu, M.; Dong, M.; Duan, Y. A mobile vision inspection system for tiny defect detection on smooth car-body surfaces based on deep ensemble learning. *Meas. Sci. Technol.* **2019**, *30*, 125905. [CrossRef]

photonics

MDPI

Article

Adaptive Freeform Optics Design and Multi-Objective Genetic Optimization for Energy-Efficient Automotive LED Headlights

Shaohui Xu [1,2], Xing Peng [1,2,*] and Ci Song [1,2]

1 College of Intelligent Science and Technology, National University of Defense Technology, Changsha 410073, China; xsh0411@nudt.edu.cn (S.X.)
2 Hunan Provincial Key Laboratory of Ultra-Precision Machining Technology, Changsha 410073, China
* Correspondence: pengxing22@nudt.edu.cn

Abstract: In addressing the design imperatives of automotive headlight miniaturization and energy conservation, this paper puts forth a design methodology for vehicle lighting systems that is predicated on free surface optics and an intelligent optimization algorithm. The establishment of the energy mapping relationship between the light source surface and the target surface is predicated on relevant performance standards. The numerical calculation is then integrated with MATLAB R2022a to obtain the free-form surface coordinate points and establish a three-dimensional model. To optimize the parameter design, a genetic algorithm is employed to fine-tune the design parameter θ_{max}, thereby attaining the optimal θ_{max} that strikes a balance between volume and luminous efficiency. The experimental results demonstrate that by integrating the optimal incidence angle into the design of the high beam and low beam, the final simulation results show that the optical efficiency of the low beam is 88.89%, and the optical efficiency of the high beam is 89.40%. This enables the automotive headlamp system to achieve a balance between volume and luminous efficiency. The free-form lamp design framework proposed in this study provides a reference for the compact design and intelligent optimization of the lamp system.

Keywords: freeform surfaces; automotive headlights; LED lights; genetic algorithms

1. Introduction

Amid the waves of globalization and rapid economic expansion, energy shortage and the escalating threat of environmental pollution have increasingly come to the fore as matters of grave concern [1]. China's "14th Five-Year Plan" proposes to achieve the goal of 100% utilization of high-efficiency lighting sources, among which LED lamps as a light source of lighting have many advantages and have become one of the safe and energy-saving lighting industries with great development potential [2–4].

Statistical data indicate that 60% of global traffic accidents occur at night, while 22% of accidents occur in conditions of poor illumination and visibility of vehicle headlights [5–7]. The quality of automobile headlights for road illumination is of significant importance to drivers, as it directly affects their ability to obtain road information and subsequently their judgment of the road environment [8]. The 2022 China Semiconductor Lighting Industry Development Blue Book, published by the CSA, indicates that LED lights have become the dominant technology for automotive lighting. In addition, LED lights offer a high degree of freedom concerning external lamps, which can be utilized to great effect in automotive styling design. This enables the creation and design of a brand-new vehicle that meets the user's personalized needs [9].

A considerable body of research has been conducted on the design of free-form mirrors for LED light sources. The design methods employed have included the grid method, the mating focal flow surface method, and the M-A method [10]. Yang et al. [11] proposed an oblique triangle energy mapping scheme and obtained a free-form reflector through geometric iterative calculation, and its headlight effect can reach 81%. Wang et al. [12] proposed a refractive design scheme for a pre-fog light lens combining an optical lens with a TIR freeform surface, and the light efficiency of the system was 76%. Zhu et al. [5] used the integrable ray mapping method to design the freeform lens of the headlamp without any other baffles or reflectors, and the energy efficiency of the optical system was more than 83.5%. Hsieh et al. [13] used a synchronous multi-surface (SMS) design method to construct a high-focus multi-surface reflector, which reduced the energy consumption of the headlight system by 70% and the optical loss by 30%. Alfred et al. [14] propose a method to use the MF concept with a combination of real and virtual light sources to achieve high-efficiency LED lighting in the automotive headlamp design. The efficiency gain was shown to improve to a significant value of 51.2% using our method while also keeping the optical design compact.

These are designed to better meet the requirements of the standard. However, little research has been performed on the initial parameters for the final mapping of optical efficiency and volume. Reverse design techniques are also emerging for automotive lamp design, but have not been widely applied due to the complexity of the algorithms and the time required to accumulate the relevant designs. In this paper, starting from the traditional light mapping method, we propose to use the control function to control the distribution of light intensity on the target surface to design the free-form surface, and to optimize the design parameters by genetic algorithms, so that the reflector can strike a balance between volume and luminous efficiency. The structure of this paper is summarized in the following section. Section 2 presents the design and evaluation of a free-form surface for the low-beam highlight, using the grid method as the design method and incorporating a control function to regulate the illuminance of the target surface. Section 3 presents the design and evaluation of the high-beam headlight. Section 4 describes a genetic algorithm used to optimize the lighting structure of the headlamp. Finally, Section 5 concludes this paper.

2. Design and Evaluation of Freeform-Based Low-Beam Headlight

2.1. Design Requirements of the Low-Beam Headlight

The national standards for low beams and high beams are identical, both of which are the requirements for the light distribution shape and brightness of the target face at 25 m from the light source. Figure 1 depicts the national standard GB25991-2010 [15] for low beam light distribution shape and illumination requirements. The overall requirements can be divided into two parts, as indicated by the center line [16], for the left side. The objective is to minimize the impact on the field of vision of oncoming drivers by preventing the generation of glare while ensuring that the illumination is as large as possible. Concerning the right side of the requirements, the illumination and the range of the largest possible are of paramount importance. The main purpose of this piece is to illuminate the road in front of the vehicle and to provide the driver with clear road information. Area I is the near lighting zone, which serves to enhance the illumination of the nearby. Area II and Area IV are high-brightness zones. These two zones occupy a central position in the driver's field of vision. The 50V and 50R points are located in this area, which is a requirement for the illumination of the farther away places. Therefore, it is necessary to gather the light to form a high-brightness area. Region III is the anti-glare zone, comprising the left horizontal line and the right 15° line forming the dim cut-off. In this zone, the illuminance must be maintained at a very low level, with the primary objective of preventing glare from one's headlight from affecting the driver of the oncoming vehicle [13].

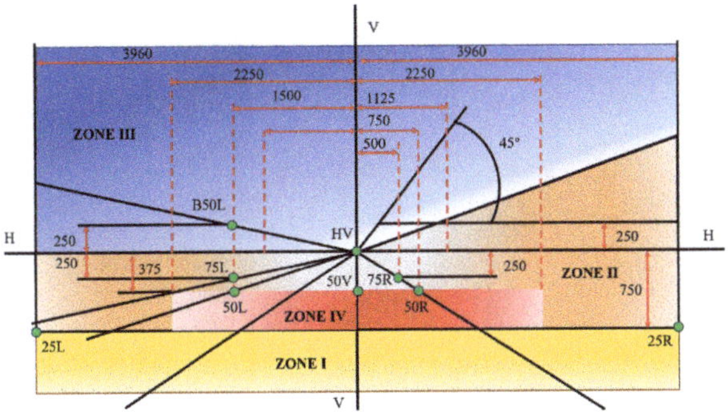

Figure 1. Lighting requirements of the low-beam headlight.

2.2. LED Light Source Energy Division

The light intensity of the light emitted from the LED headlight is approximately Lambertian, and its light intensity distribution characteristics must be considered when dividing the energy of the light source. Since the ratio of the LED light-emitting aperture to the aperture of the entire optical system is greater than 1/5, the light source can be treated as a point light source, disregarding the impact of the size of the light source on the design [17]. In consideration of the LED light distribution characteristics, the division of the light source surface is typically conducted through the use of latitude and longitude lines, as illustrated in Figure 2.

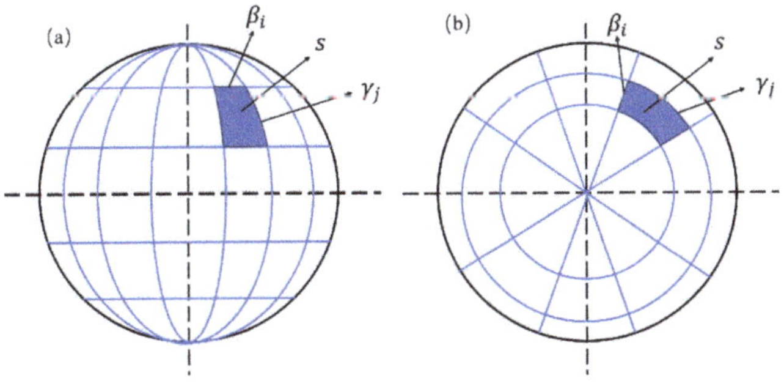

Figure 2. Schematic diagram of latitude and longitude divisions: (**a**) front view; (**b**) top view.

Under the entire light path process, the coordinate system of light propagation is set up. Firstly, the position of the light source is set as the coordinate origin, recorded as point O. The optical axis direction points to the negative direction of the Z-axis, that is, the LED headlight is downward irradiated. The X-axis represents the direction of forward motion, while the entire light mapping trajectory for the O-point is shot in the negative direction of the Z-axis, which is the reflective surface of the collision of the reflection of the X-positive direction. This results in the formation of the target spot at X = 25,000 mm. It can be assumed that any ray of light is represented by the vector $\overrightarrow{OP}$. In this context, the angle between the vector $\overrightarrow{OP}_{(i,j)}$ and the negative half-axis of the X-axis is denoted by θ_i. Furthermore, the angle denoted by γ_j is the angle between the plane formed by the vector

$\overrightarrow{OP}_{(i,j)}$ and the negative half-axis of the X-axis and the XOY plane. The combination of the angles θ_i and γ_j forms the angle of any outgoing light source, as illustrated in Figure 3.

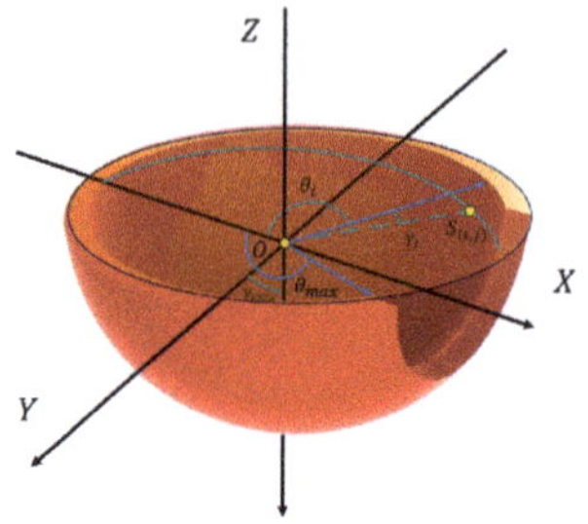

Figure 3. Schematic diagram of light source division.

The LED light source is divided into M × N parts, ensuring that each part has the same luminous flux. The luminous intensity distribution $I = I_0 \sin\theta \sin\gamma$ can be obtained according to the form of division. The total luminous flux emitted by the light source is:

$$\Phi = \int_0^{\gamma_{max}} d\gamma \int_0^{\theta_{max}} I\sin\theta d\theta \tag{1}$$

The energy of either source can be described as follows:

$$d\Phi = \int_{\gamma_j}^{\gamma_{j+1}} d\gamma \int_{\theta_i}^{\theta_{i+1}} Isin\theta d\theta = \frac{\Phi}{MN} \tag{2}$$

$$d\Phi = \int_{\gamma_j}^{\gamma_{j+1}} sin\gamma d\gamma \int_{\theta_i}^{\theta_{i+1}} sin^2\theta d\theta \tag{3}$$

Once the division is complete, the rays can be represented as vectors. For example, the ray OP(i, j) can be represented as $(-\cos\theta, -\cos\theta\sin\gamma, -\sin\gamma)$.

2.3. The Division of the Low Beam Target Surface

From the preceding section of the national standard for low-beam headlights, it can be observed that the principal challenges in the design of low beam free-form mirrors are the low brightness on the left side, the high brightness on the right side, and the dim cut-off line. Accordingly, the objective of this paper is to propose a novel design concept that separates the reflector into two distinct components, as illustrated in Figure 4. This approach entails dividing the reflector and the target light distribution surface from the vertical center line into two separate sections, with the reflector R1 and rectangular light distribution spot T1 situated on the left and the reflector R2 and trapezoidal light distribution spot T2 on the right.

Firstly, the left target surface, as shown in Figure 5, must be analyzed with a particular focus on the highlight area. This area will be divided into equal areas according to the gamma angle. The division line is called the control line. By controlling the distribution of points on the control line, it is possible to achieve the distribution of light spot energy on the target surface. The value of γ_j in Figure 5 is equivalent to that of γ_j in Figure 3. The grid is divided by the same starting point, the angle of incidence, γ_j, in the range of 0 to γ_{max}, with one side of the grid line angle divided by the control function. The length of each grid line is divided by the control function. The target surface corresponds to the γ_j of the light source surface, while the target surface of the control function corresponds to the light source surface of the θ_i.

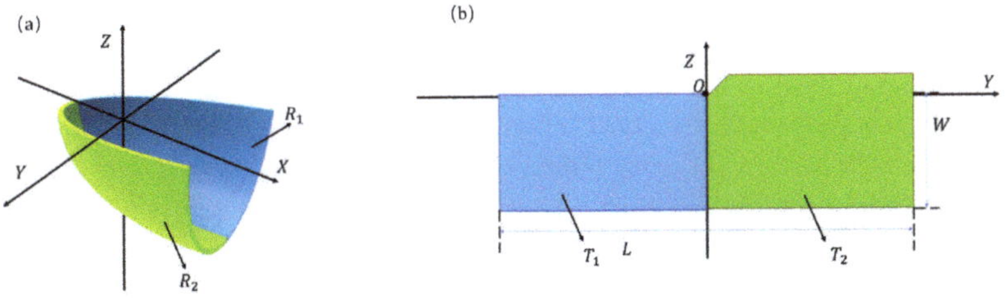

Figure 4. The design scheme: (**a**) the separation of mirrors on the left and right; and (**b**) the separation of target surfaces on the left and right.

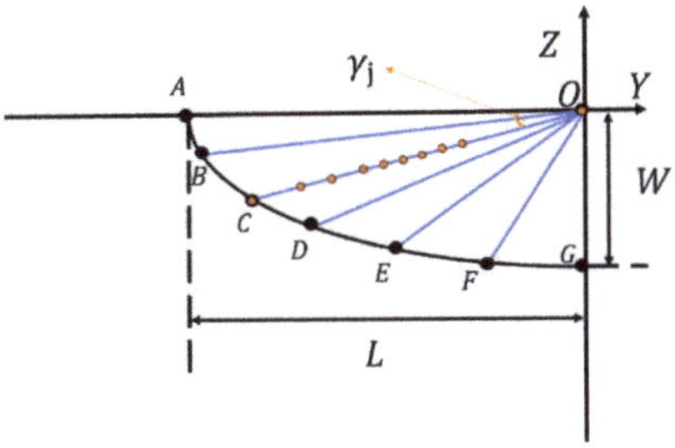

Figure 5. Energy analysis on the left side of the low-beam headlight.

Concerning the control line OC, it is possible to take 10 sampling points on the control line as control points. The distribution of points on the entire control line can be controlled by regulating the distribution of these 10 points, thus achieving control of the light intensity distribution. Under the established standards, the left side of the high-brightness area must be situated in the middle and lower part of the target surface to fulfill the requirements of 75L and 50L in terms of light intensity. Additionally, the center of the target surface for the 25 L also requires brightness. Consequently, it is essential to concentrate the control points in the central and lower regions to fulfill the requirements of 75L and 25L. Furthermore, the closer to the C point, the fewer the control points should be to prevent the emergence of high brightness in the I area, which does not meet the design requirements. The ratio of the distance of the control point from the origin and the length of the control line is referred to as the Proportions. The order of the control points is used to determine the curve fitting, which allows the control function of the control point to be obtained, as shown in Figure 6.

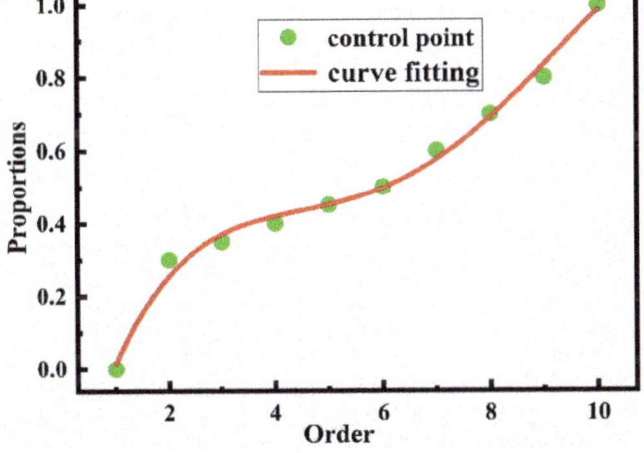

Figure 6. Fitting the control function for the left side of the low-beam.

Secondly, the right side of the analysis will be divided into equal areas following the γ angle. As illustrated in Figure 7, the design specifications at this stage primarily concern the creation of a high-brightness area in the upper region and the implementation of a 45° level of dark and light cut-off lines. This is exemplified by the HV-H1-H2-H4 line, which necessitates the control line at the point to be sparse in front of the front end, before becoming dense in the middle and sparse again at the back end. This approach serves to collect the control points.

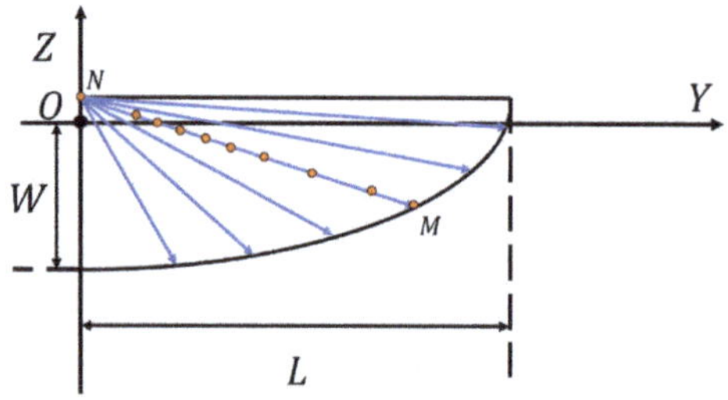

Figure 7. Energy analysis on the right side of the low-beam headlight.

Concerning the control line MN, it is possible to take 10 sampling points on the control line as control points. The distribution of points on the control line as a whole can be controlled by regulating the distribution of these 10 points, thus achieving control of the light intensity distribution. Following the established standards, the left side of the high-brightness area must be situated in the middle and upper parts of the 75R and 50R to fulfill the requisite specifications for light intensity. Additionally, the brightness of the 25L must be maintained away from the center of the target surface. Consequently, it is essential to concentrate the control points in the upper center to achieve the appearance of the 45° cut-off line, as well as to meet the requirements of the 75L and 25L high luminance. Furthermore, the closer the control points are to the M point, the sparser they are, to avoid the appearance of high luminance in the I area, which would make it impossible to satisfy the design requirements. The control points are then collected and fitted to obtain the desired control function, as presented in Figure 8.

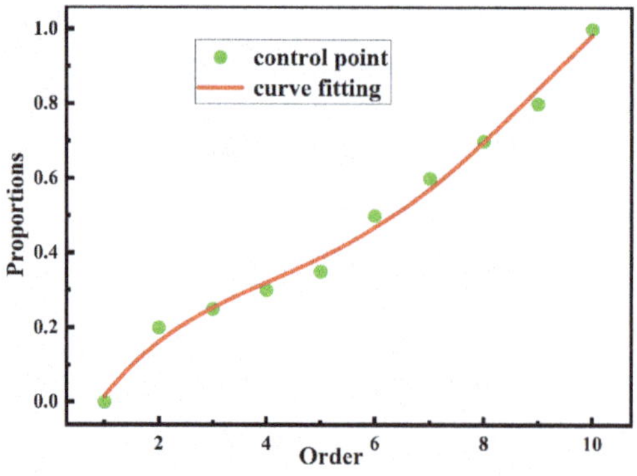

Figure 8. Fitting the control function for the right side of the low-beam.

2.4. Calculation of Free-Form Surfaces for Low-Beam Headlight

In the preceding section, the light source is divided into equal energies, and the target surface is also divided into energies [18]. Finally, the light intensity distribution is determined by a control function. The mapping path for the light–mirror–target surface of the right half of the free-form reflector is shown in Figure 9. This allows for the free-form surface calculation to be carried out.

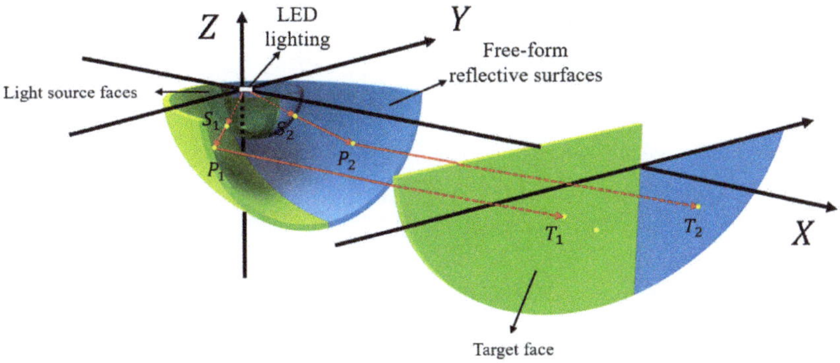

Figure 9. Schematic diagram of the low-beam light mapping path.

The calculation of the free-form surface is based on the modeling of the coordinates of each point on the surface. Therefore, it is necessary to obtain the positions of the points on the surface model in space before the calculation. As illustrated in Figure 10, the coordinates of each point on the free-form surface are calculated iteratively using the values of θ_i and γ_j. To illustrate, consider the calculation of one of the construction curves L_1. A certain vector of light emitted from a light source is designated as I1, and the vector of light directed to the target surface after reflection is designated as O_1.

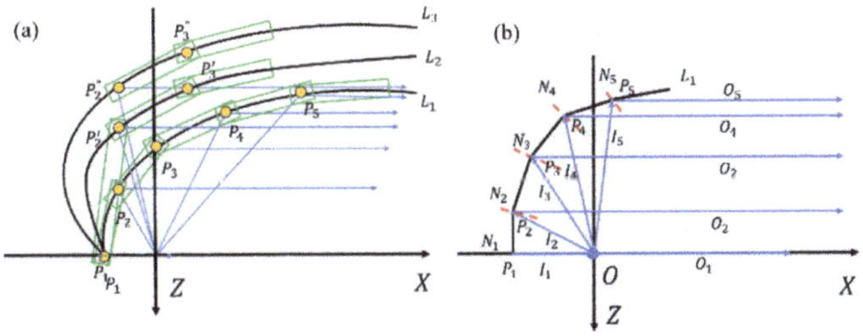

Figure 10. Schematic diagram of the methodology employed for the calculation of a freeform surface: (**a**) the construction of the surface itself; and (**b**) the construction of the curve.

If the light source is situated at the origin O of the three-dimensional coordinate system, and the initial point of the surface construction is P_1, the vectors $\vec{I}$ and $\vec{O}$ can be calculated based on the energy division of the light source as follows:

$$\vec{I} = \left(\sin\theta_i \cdot \sin\gamma_j,\ \sin\theta_i \cdot \cos\gamma_j,\ \cos\theta_i\right) \tag{4}$$

$$\vec{O} = \frac{\left(T_x - f(T_x), T_y - f(T_y), D - f(T_z)\right)}{\sqrt{\left((T_x - f(T_x))^2 + \left(T_y - f(T_y)\right)^2 + (D - f(T_z))^2\right)}} \tag{5}$$

The sum of the vectors $\vec{I_1}$ and $\vec{O_1}$ can thus be calculated from the above two formulas, with the inclusion of Fresnel's law to calculate the normal vector $\vec{N_1}$ of P_1 under the plane formed by the curve of L_1. This is as follows:

$$\left[2 - 2\cdot\left(\vec{I_1}\cdot\vec{O_1}\right)\right]^{\frac{1}{2}}\cdot\vec{N} = \vec{O_1} - \vec{I_1} \tag{6}$$

Once the normal vector, designated as $\vec{N_1}$, has been obtained at point P_1, it can be assumed that the tangent vector of $\vec{N_1}$ is T_1. The prolongation vector, designated as $\vec{I_2}$, intersects with the tangent vector $\vec{T_1}$ at point P_2, which is the next point on the curve of the free-form surface L_1.

$$\vec{OP_2} - \vec{OP_1} = \vec{T_1} \tag{7}$$

The ratio of vector $\vec{OP_2}$ to vector $\vec{I_2}$ is defined as x.

$$\vec{OP_2} = x\cdot\vec{I_2} \tag{8}$$

The vector–perpendicular relationship can be demonstrated by the following equation:

$$\vec{T_1}\cdot\vec{N_1} = 0 \tag{9}$$

The solution to the three equations presented above can be determined by linking them together.

$$x = \frac{\vec{OP_1}\cdot\vec{N_1}}{\vec{I_2}\cdot\vec{N_1}} \tag{10}$$

Substitution of the x-value into the formula of $\vec{OP_2}$ allows the coordinates of point P2 to be determined, and this process can be repeated for all points on the L1 curve. By varying the angle of γ_j, the value of the subsequent construction curve can be determined, and the coordinates of all points on the free-form surface can be obtained once all construction curves have been calculated, as presented in Figure 11.

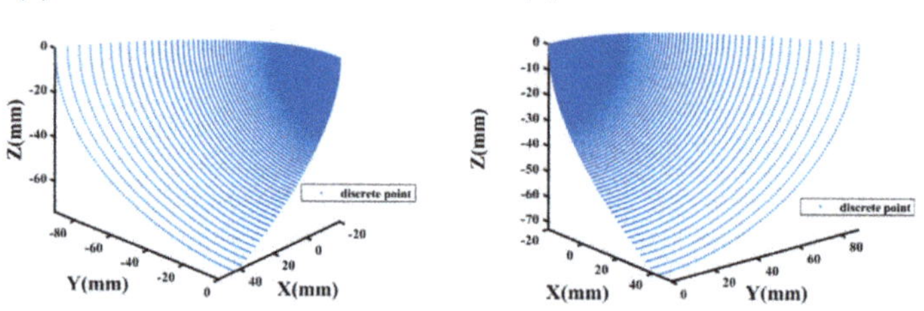

Figure 11. Calculation results of the free-form discrete points: (**a**) discrete points on the left side of the reflector; (**b**) discrete points on the right side of the reflector.

This construction method allows the entire structure curve to be derived directly from a single starting point, thus circumventing the necessity to utilize the primary curve. In the case of such surfaces in three-dimensional space, it is typically necessary to construct a primary curve and use it as an initial point to construct a secondary curve. However, this method can easily result in the accumulation of errors in the calculation, which may subsequently lead to points behind the calculation deviating more and more from the real situation. The design method presented in this paper has the potential to reduce the accumulation of errors and optimize the design efficiency.

2.5. Modeling and Simulation of the Low-Beam Headlight

In the preceding section, the coordinates of the discrete points of the low beam reflector were calculated in MATLAB R2022a. Subsequently, surface fitting was performed in UG, resulting in a fitted physical model with a length of 179.08 mm, a width of 62.68 mm, and a height of 72.91 mm, as illustrated in Figure 12.

Figure 12. The designed low-beam headlight model.

The constructed model is imported into ZEMAX 2023R2 for light tracing simulation. The light source is the same as the high beam OSRAM OSLON® Black Flat (https://ams-osram.com/ (accessed on 16 February 2025)), with a luminous flux of 365 lm. As illustrated in Figure 13, the light source is situated at the origin, while the detector is positioned 25 m away from the origin. The simulated illuminance diagram is presented in Figure 14, while the simulated illuminance values are displayed in Table 1.

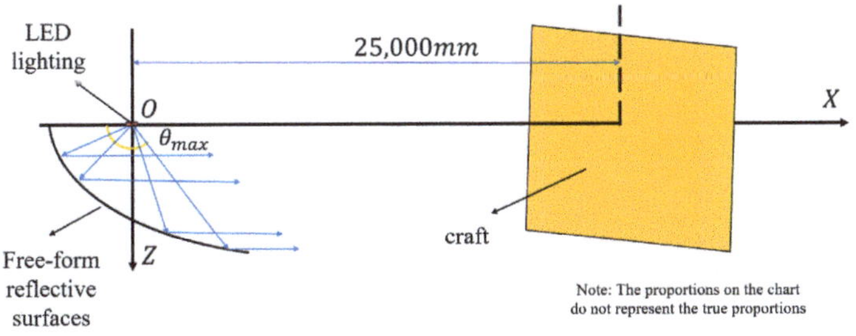

Figure 13. Schematic diagram of the simulated optical system.

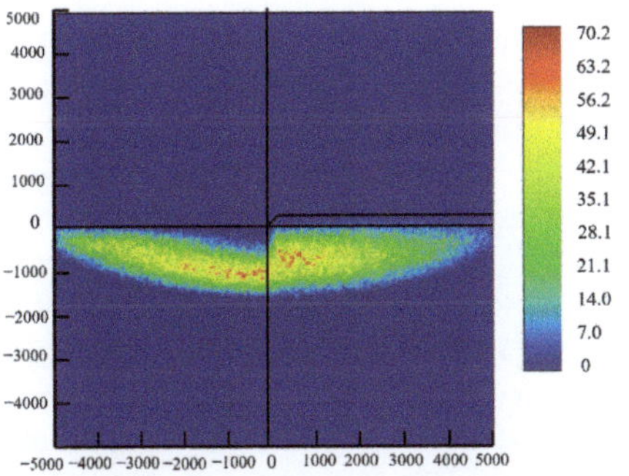

Figure 14. Simulation result of the low-beam headlight.

Table 1. The illuminance values list of the low-beam headlight.

Test Points or Areas	Required Illuminance (lx)	Simulated Illuminance (lx)	Fulfillment of Requirements
HV	≤0.7	0.57	Yes
B50L	≤0.4	0	Yes
75R	≥12	32.16	Yes
75L	≤12	10.21	Yes
50L	≤15	13.32	Yes
50R	≥12	48.43	Yes
50V	≥6	25.71	Yes
25L	≥2	10.31	Yes
25R	≥2	7.08	Yes
III area	≤0.7	≤0.14	Yes
VI areas	≥3	≥6.85	Yes
I areas	≤2 × 50R	≤54.33	Yes

The simulation results indicate that the proposed design scheme is capable of meeting the design requirements and achieving superior light uniformity. In the entirety of the optical path, the light is divided into three distinct categories, as illustrated in Figure 15. These include the light that can be reflected normally, the light that is reflected twice by the reflector, and the light that is not reflected by the reflector. Among these, the light that reaches the designated area only after undergoing normal reflection is considered effective, while the remaining two categories are deemed ineffective. The energy efficiency of this optical system is calculated as follows:

$$\eta = \frac{\Phi_T}{\Phi_S} \tag{11}$$

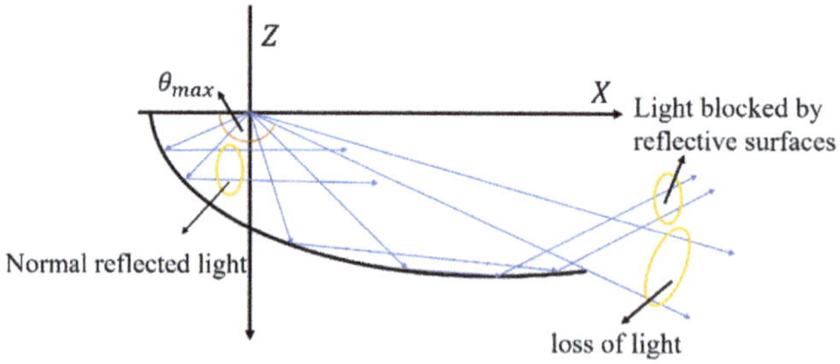

Figure 15. Schematic diagram of energy loss.

The luminous flux received by the target surface, denoted by Φ_T, and the luminous flux emitted by the light source, denoted by Φ_S, are the two variables of interest in this context. The calculated energy utilization efficiency is 80.59%, with 19.41% of the energy dissipated throughout the process. Concerning the automobile, this component of dissipated energy can enhance the vehicles near illuminance. However, this is not a viable solution for the high beam.

The light will propagate along the X-axis in a positive direction throughout the optical system, and the reflector must remain unobscured at all times. The value of the θ_{max} angle has a significant impact on the optical efficiency of the system. When θ_{max} is small, there is a limitation on the amount of light that can be reflected to the X-axis in a positive direction. This results in a reduction in optical efficiency. Conversely, when θmax becomes larger, the mirror volume increases rapidly, which blocks the reflected light. Consequently, it is essential to strike a balance between size and optical efficiency and to assign an appropriate value to the θ_{max} angle. The angle θ_{max} is 120°. The optimization will be conducted subsequently to identify the smallest possible size with the greatest possible luminous efficiency.

3. Freeform-Based High-Beam Headlight Design and Evaluation

3.1. Design Requirements of the High-Beam Headlight

The requirements of the national standard for high beams are primarily concerned with the brightness of the light distribution screen 25 m away from the headlight and the perpendicular direction of the car driving. The standard is relatively lenient, with only five test points and intervals specified, as shown in Figure 16 and Table 2.

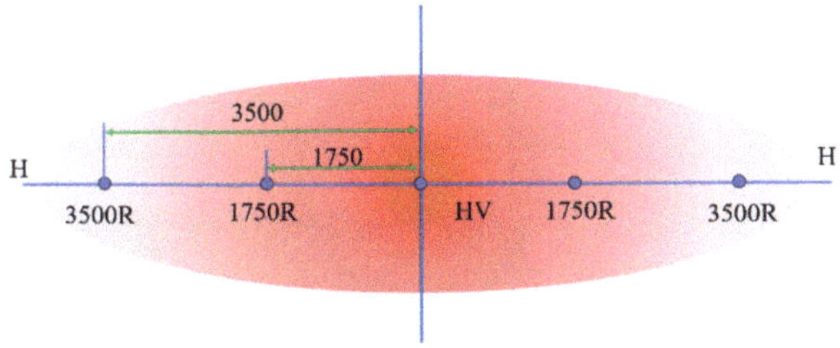

Figure 16. Lighting requirements of the high-beam headlight.

Table 2. Light distribution requirements of the high-beam headlight.

Test Points or Areas	Required Illuminance (lx)
E_{max}	≥48 and ≤250
HV Point	≥0.8 E_{max}
1125L to HV and HV to 1125R	≥24
2250L to HV and HV to 2250R	≥6

3.2. High-Beam Reflector Design

The specifications for vehicle front-illuminated high beams only require that the illuminance on the horizontal plane be met, but do not stipulate the size or shape of the light spot hitting the object surface [19]. To achieve a more reasonable distribution of light and a superior lighting effect, it is necessary to ensure that the target surface design spot division and the light source latitude and longitude of the division correspond. To align the light source division with the target surface division, it is necessary to modify the traditional target surface division. The new target surface division form is illustrated in Figure 17.

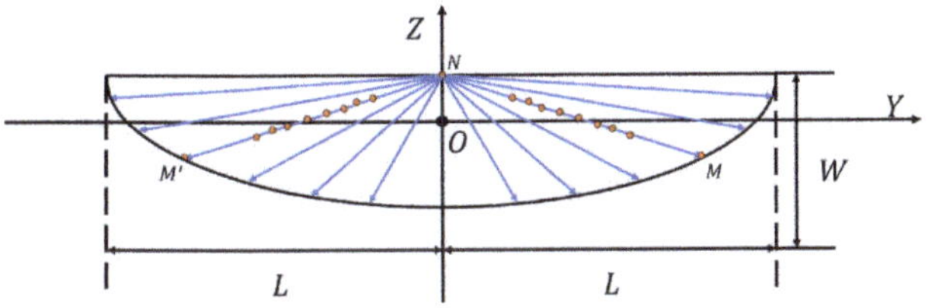

Figure 17. Schematic diagram of the target surface division.

To illustrate, the control line OM may be considered a case in point. Ten sampling points on the control line may be taken as control points, and the distribution of points on

4. Genetic Algorithm-Based Intelligent Optimization Method for Headlight Optical Structures

4.1. The Optimization Model Establishment

One of the most significant factors influencing the optical efficiency of the headlight reflector is the maximum angle of reflection, θ_{max}. Increasing θ_{max} will result in a greater proportion of reflected light; however, this will also result in a proportion of the light that has already been reflected being blocked. Consequently, it is not possible to simply increase θ_{max} to improve optical efficiency. Instead, it is necessary to consider the reflector's effect on the reflected light. Concurrently, an increase in the value of θ_{max} results in a rapid expansion of the reflector volume, which is contrary to the design objective of a compact configuration. Consequently, it is essential to optimize and enhance the reflector's optical efficiency and volume simultaneously.

To illustrate this point, consider the high beam mirror. The relationship between the light effect and the volume is determined by simulating the left half of the high beam mirror. The ensuing simulation results are displayed in the accompanying Figure 22:

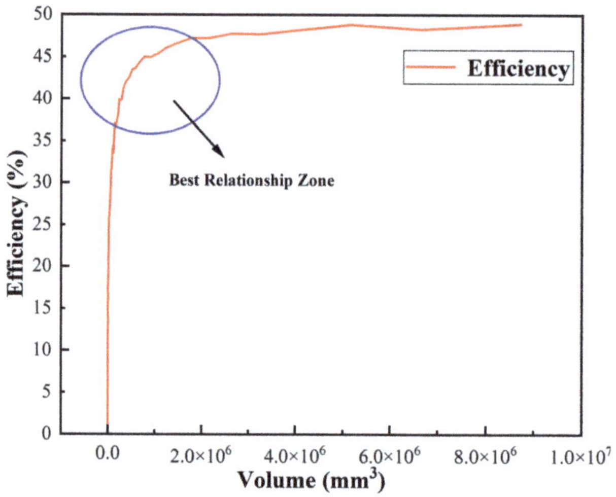

Figure 22. Volume–efficiency.

It is evident that an increase in volume at the initial stage results in a rapid rise in light efficiency. However, as the volume reaches 2×10^5 mm, the rate of growth in light efficiency begins to decelerate. Subsequently, when the volume reaches 2×10^6 mm, the growth of light efficiency becomes negligible. Consequently, the volume range between 2×10^5 mm and 2×10^6 mm can be considered the optimal region. Consequently, identifying the appropriate θ_{max} value is imperative to ascertain the optimal volume and light efficiency within the designated optimal corresponding region.

The range of θ_{max} can be 0° to 360°, but when $\theta_{max} \geq \pi/2$, the reflector will only appear to block the reflected rays; when $\theta_{max} \geq 5\pi/6$, the volume of the reflector becomes dramatically larger, and it does not satisfy the design requirement of a small volume. Consequently, the maximum angle of incidence, θ_{max}, is constrained to the range of 0° to $5\pi/6$. Based on the value of θ_{max}, a front-illuminated lighting system is established, and then the optical simulation is carried out in TracePro to obtain a series of data, including a simulated light map. To process the data, it is necessary to utilize an evaluation function to establish a unified index. The establishment of the simulation evaluation function must consider the impact of θ_{max} on optical efficiency and volume, as well as the impact of light efficiency and volume on the evaluation function of weight. This is achieved by applying the principle of normalization to establish the evaluation function.

$$f_i = \omega_1 \frac{\eta_i - \eta_{min}}{\eta_{max} - \eta_{min}} - \omega_2 \frac{V_i - V_{min}}{V_{max} - V_{min}} \tag{12}$$

In the formula, the concept of normalization is employed for both η and V. Following the normalization of these variables, the evaluation weights, ω_1 and ω_2, are calculated. Their size is subsequently adjusted according to the specific design requirements. Finally, the system is optimized to identify the value of θ_{max} in the case of the maximum value of f_i.

The optimization of free-form surfaces is a subcategory of optimal path optimization. Several optimal path algorithms can be applied in the optimization process of the design of the front-illuminated lighting science system, including the genetic algorithm, annealing algorithm, and particle swarm algorithm [20]. In the optimization process of this system, it is necessary to perform direct operations on the free surface. There is no qualification of derivation or function continuity, which makes it impossible to obtain a definitive rule to guide the search space of optimization. After comprehensive consideration, the genetic algorithm was selected as the optimal method for optimization [21]. This approach allows us to identify the optimal solution without having to determine the optimal result, as the nature of the optimization process is not always clear. Instead, we focus on eliminating suboptimal solutions and iterating until we reach the optimal solution without easily falling into a local optimum. Consequently, genetic algorithms are an optimal choice for optimizing free-form surfaces, as they facilitate the identification of the optimal θ_{max} that balances light efficiency and volume [22].

4.2. Optimization Process for Headlight Optical Structures

The previously mentioned design method for the free-form reflector for the low and high-beam headlights was employed to generate a point cloud in MATLAB R2022a. The boundaries of the point cloud are identified to ascertain the extent of the space occupied by the mirror, thereby enabling the calculation of its volume. Concurrently, to iterate the genetic algorithm, it is necessary to automate the modeling process. This involves selecting the point cloud to be processed in MATLAB R2022a, after which the points are processed to create an STL solid file.

The point cloud data obtained in the previous section were initially reduced to a two-dimensional state to obtain the projection of the point cloud image onto the YZ coordinate plane, as shown in Figure 23a. The points projected into the YZ surface are then processed using the Delaunay algorithm [23] to obtain a triangulation connection list of the point cloud within the YZ projection plane, as shown in Figure 23b. This list contains information on the number of triangulation surfaces formed by the dissections of this point cloud and the points corresponding to the vertices of each triangulation surface. The 2D triangulation connection list for the mirror-free form surface is identical to that of the 3D case, thus enabling mapping to the 3D case. The triangulated connections are mapped to the coordinates of their associated vertices in three dimensions, as presented in Figure 23c, and the triangulated connections are then written to a binary STL file with the coordinates of the vertices. This concludes the CAD modeling process, which was conducted using MATLAB to process the point cloud data.

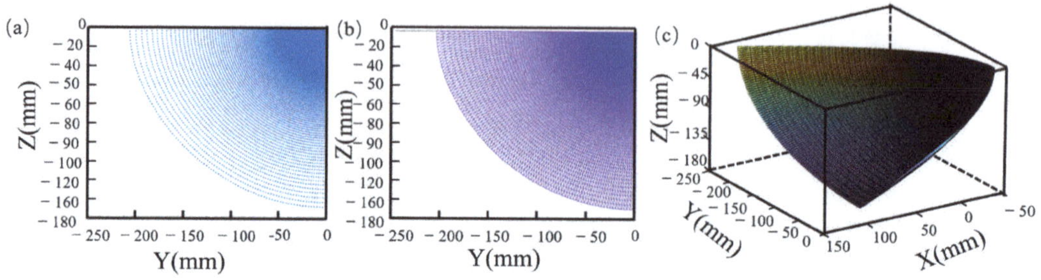

Figure 23. The construction process of an STL solid model: (**a**) YOZ 2D point cloud, (**b**) triangular sectioning, and (**c**) mesh construction.

The STL solid model is constructed and then the ZEMAX kernel is invoked with MATLAB to simulate the system using a sequence-free model. The relevant parameters of the LED light source were set at a distance of 25,000 mm from the light source in the location of the rectangular detector, which was 10,000 mm long and 8000 mm wide. This detector was imported into the established STL solid file and given the material data of the reflector. The ray tracing was then executed to complete the simulation process. Finally, the collected pixel information of the detector was used to derive the optical efficiency using MATLAB R2022a. The collected pixel information from the detector is used to derive the optical efficiency.

To optimize the optical efficiency and volume of the free-form surface, the genotype θ_{max} is employed as the screening criterion, with the free-form surface individual under this genotype being obtained through iterative computation and MATLAB modeling. The evaluation function, calculated following the normalization of volume and light efficiency, serves as the fitness function in the genetic algorithm, enabling the determination of the fitness of the individual. A roulette algorithm was employed to identify free-form individuals with low light efficiency. It is still possible to select surfaces with low fitness using roulette, although the probability of this occurring is low and the genes are not completely discarded. Consequently, the most exemplary members of the preceding population are combined with a certain probability of forming a new population through a selection operation, which results in the reproduction of offspring [24].

$$P_i = \frac{f_i}{\sum\limits_{k=1}^{M} f_k} \tag{13}$$

Accordingly, the optimal individuals from the previous generation are combined with a certain probability of forming a new generation, to achieve the reproduction of offspring.

As the genetic algorithm iterates to identify the optimal value, the iteration process cannot be expressed as a function. Consequently, the value of the iteration process is not clear. The termination of the evolutionary conditions is indicated by the termination of the genetic generation, as shown in Figure 24.

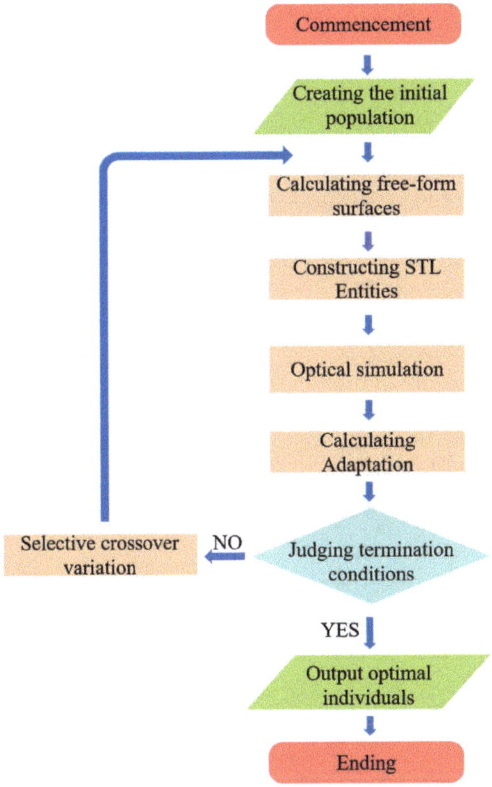

Figure 24. Optimization flow chart of the headlight structure.

4.3. Results and Analyses

The optimization can be attributed to single-parameter genetic algorithm optimization. Referring to related literature, the number of individuals in the population of Jiang is generally set to 10–25, and the number of genetic generations is set to 10–20 times [25]. Given the intricacy of the model, the number of individuals in the population is set to 20, and the number of genetic generations is set to 15 in the simulation of this paper. To facilitate the genetic algorithm iteration of ZEMAX, a communication channel was established between MATLAB and ZEMAX, and the MATLAB code was written to implement this iteration.

According to the results obtained after the iteration, it can be observed that when the number of iterations reaches 14, the fitness level attains its maximum value in Figure 25. At this juncture, it can be posited that the parameter of the population aggregation prior to and following the 14th iteration constitutes the optimal solution for the specified objective. The results from the fifteenth iteration will be examined. A thorough examination of the error of the simulation reveals that in the population aggregated at $\theta_{max} = 2.2$ in the 15th generation, the evaluation function of the individuals is situated in the uppermost interval. Based on Figure 26, the fitness level is recorded as 0.8. Subsequent to incorporating the value of $\theta_{max} = 2.2$ into the computation of θ_{max}, as detailed in Sections 2 and 4, the optimal values of the high and low beam free-form mirrors are determined, as illustrated in Figure 27.

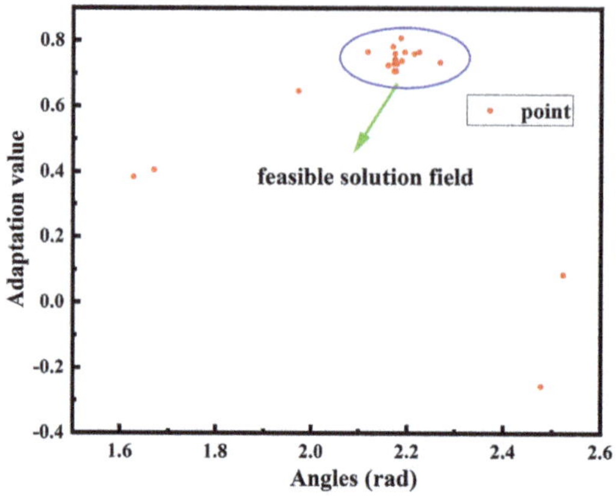

Figure 25. The distribution of individuals in the fifteenth generation.

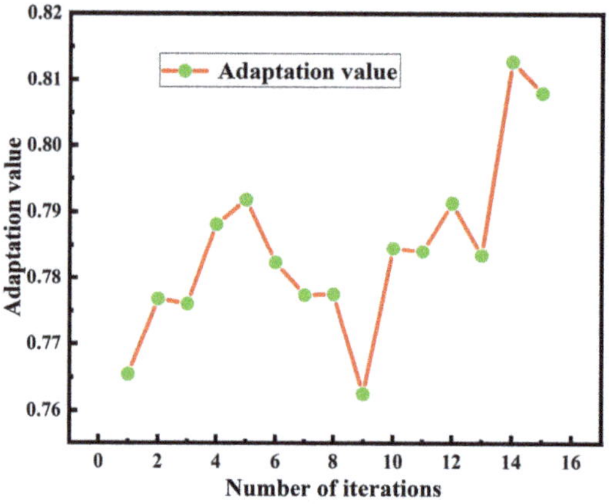

Figure 26. Variation of the best solution per generation.

Figure 27. The optimized high-beam headlight model.

Then, the optical tracing simulation was carried out in ZEMAX to obtain its simulated light Figure 28, and the light values of each test point and test area as shown in Table 4.

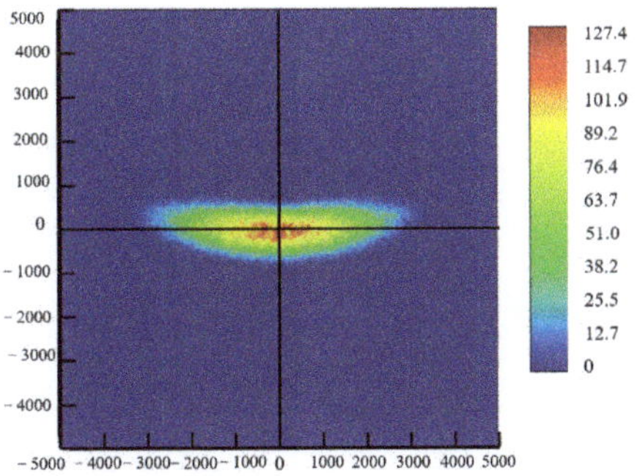

Figure 28. The optimized lighting map of the high-beam headlight.

Table 4. The results of the optimized high-beam headlight.

Test Points or Areas	Required Illuminance (lx)	Simulated Illuminance (lx)
E_{max}	$\geq$48 and $\leq$250	127.4
HV Point	$\geq 0.8\ E_{max}$	110.4
1125L to HV and HV to 1125R	$\geq$24	87.90
2250L to HV and HV to 2250R	$\geq$6	38.62

The E_{max} values are within the range of 48 lx to 250 lx, as stipulated by the standard. Furthermore, the illuminance at point HV exceeds 0.8 Emax. The minimum illuminance value at HV points 1125L and 1125R is 87.90 lx, while the minimum illuminance value at HV points 2250L and 2250R is 38.62 lx. These values satisfy the design requirements and are comparable with the values of each test point or area before optimization. The illuminance values of each test point or area have been enhanced in comparison to those observed before optimization. The luminous efficacy of the aforementioned device is 80.69%, and it is capable of achieving better optical uniformity, thereby avoiding the occurrence of high illuminance in the center and improving the global illuminance simultaneously. The length is 218.94 mm, the width is 101.14 mm, the height is 90.82 mm, and the optical efficiency is 89.40%. In comparison to the high beam obtained in Section 3, the length has been extended by 61.16 mm, the width by 40.79 mm, and the height by 20.93 mm, while the optical efficiency has been improved by 8.71%.

The shape of the low-beam is shown in Figure 29.

Figure 29. The optimized low-beam headlight model.

Subsequently, an optical tracing simulation is conducted in ZEMAX, as presented in Figure 30. The light values of each test point and test area are presented in Table 5.

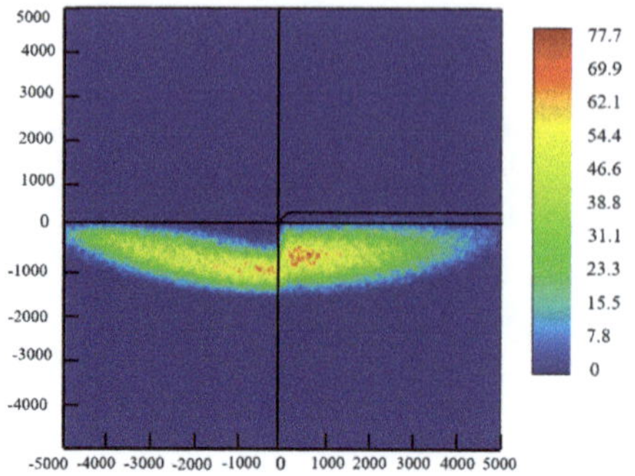

Figure 30. The optimized lighting map of the low-beam headlight.

Table 5. The simulation results of the low-beam headlight.

Test Points or Areas	Simulated Illuminance (lx)	Fulfillment of Requirements
HV	0.5	Yes
B50L	0	Yes
75R	42.64	Yes
75L	10.73	Yes
50L	14.83	Yes
50R	49.07	Yes
50V	36.72	Yes
25L	10.94	Yes
25R	9.63	Yes
III area	$\leq$0.46	Yes
VI areas	$\geq$7.95	Yes
I areas	$\leq$69.88	Yes

The dimensions of the headlight are as follows: length 255.72 mm, width 105.81 mm, height 96.04 mm. The optical efficiency is 88.89%. In comparison with the low-beam headlight previously described, the length was increased by 76.64 mm, the width by 43.13 mm, and the height by 23.13 mm, while the optical efficiency was improved by 8.30%.

It can be observed that the design of the low beam and high beam conforms to the specifications outlined in the national standard. To achieve a compact size while maintaining high optical efficiency, the design of this paper's headlight outperforms that of conventional automotive headlights.

4.4. Manufacturing Process of Automotive Freeform Mirrors

The manufacturing of automotive headlight mirrors comprises three primary processes: injection molding, painting, and aluminizing. Compared to the manufacturing of conventional lamps, free-form lamp manufacturing entails increased costs associated with mold fabrication. Conversely, injection molding facilitates mass production, thereby enhancing the efficiency and cost-effectiveness of automotive headlamp manufacturing. The current processing of free-form surfaces primarily utilizes ultra-precision turning, and the polishing of the molds post-SPDT cutting necessitates the implementation of computer-controlled optical surface polishing (CCOS) and magnetorheological polishing (MRF) to enhance the surface finish [26,27].

Following injection-molding, the product surface is coated with primer to meet the stipulated surface finish requirements. Aluminizing fulfills the reflectivity requirement. The purpose of aluminizing is to improve the light reflection effect and decorative effect, which is 90% for aluminum and 90% for silver. Aluminum is a relatively economical and widely used material. The quality of the aluminized plate is a critical factor in determining light reflection [28].

5. Conclusions

In this paper, we initiate the study by employing the grid mapping method. We then introduce a control function that governs the illumination of the target surface. We subsequently complete the design and evaluation of the low beam and free-form surface. Following this, we establish an optimization model and optimize the design parameters using a genetic algorithm. This allows the reflector to achieve a balance between volume and luminous efficacy. The optimal θ_{max} is incorporated into the design of the low and high beams, and the final results are simulated. The comparison of the before and after design is shown in Table 6, which demonstrates that according to our proposed method, the volume can be controlled within a reasonable range while ensuring high efficiency. In light of the accelerated development of new energy vehicles, there will be an increase in the use of LED lamps, and the design of LED lamps will continue to evolve. This paper proposes a free curved front lighting design and optimization method that can take into account optical efficiency and volume. This method better achieves uniform illumination while meeting design requirements and providing a reference for the uniformity of the light.

Table 6. Comparison of results.

	Before Optimization		After Optimization	
	High Beam	**Low Beam**	**High Beam**	**Low Beam**
θ_{max} (rad)	2.09		2.2	
dimension (mm)	157.78 × 60.35 × 69.89	179.08 × 62.08 × 72.91	218.94 × 101.14 × 90.82	255.72 × 105.81 × 96.04
efficiency (%)	80.69	80.59	89.40	88.89

Author Contributions: S.X.: Methodology, Software, Writing—Original draft. C.S.: Conceptualization, Software, Writing—original draft. X.P.: Writing—review and editing, Supervision. All authors have read and agreed to the published version of the manuscript.

Funding: This work is supported by the National Natural Science Foundation of China (Grant No. 52305594), and the Natural Science Foundation of Hunan Province (Grant. No. 2024JJ6460).

Data Availability Statement: Data are contained within the article.

Conflicts of Interest: The authors declare no conflict of interest.

References

1. Hu, C.C.; Zheng, Y.J.; Hsu, Y.F.; Ye, Z.T. Design and Application of Liquid Silicone Rubber Light Guide in Compact Automotive Headlamps. *Int. J. Optomechatron.* **2024**, *18*, 2343407. [CrossRef]
2. Qin, R.S. New Trends in Accounting Reform and Development in the '14th Five-Year Plan' Period. *XinLiCai* **2021**, *9*, 12–17.
3. Zhang, H.; Liu, D.; Wei, Y.; Wang, H. Asymmetric double freeform surface lens for integrated LED automobile headlamp. *Micromachines* **2021**, *12*, 663. [CrossRef] [PubMed]
4. Liu, L.; Lin, L. Design of a new pixel LED automobile headlamp. In Proceedings of the 2021 International Conference on Optical Instruments and Technology: Optical Systems, Optoelectronic Instruments, Novel Display, and Imaging Technology, Online, 8–10 April 2022; pp. 140–146.
5. Zhu, Z.; Wei, S.; Liu, R.; Hong, Z.; Zheng, Z.; Fan, Z.; Ma, D. Freeform surface design for high-efficient LED low-beam headlamp lens. *Opt. Commun.* **2020**, *477*, 126269. [CrossRef]
6. Siddula, S. Design of Modern Technology Lighting System for Automobiles Check for updates. *Emerg. Technol. Sustain. Dev. Sel. Proc. EGTET* **2023**, *2022*, 1061.
7. Zeng, Y. Free-form Automotive Laser Headlight Lighting System Based on Free-form Surface Design. Master's Thesis, Changchun University of Science and Technology, Changchun, China, 2021.

8. Xie, S.S. Design and Performance Analysis of Automobile Headlamp Based on Light-Emitting Diode. *J. Nanoelectron. Optoelectron.* **2023**, *18*, 1203–1210. [CrossRef]
9. Bielawny, A.; Schupp, T.; Neumann, C. Automotive lighting continues to evolve. *Opt. Photonics News* **2016**, *27*, 36–43. [CrossRef]
10. Feng, Z.X.; Froese, B.D.; Liang, R.G.; Cheng, D.; Wang, Y. Simplified freeform optics design for complicated laser beam shaping. *Appl. Opt.* **2017**, *56*, 9308–9314. [CrossRef] [PubMed]
11. Yang, Y.S.; Qiu, D.; Zeng, Y.; Li, R.; Duan, W.; Fan, R. Design of a reflective LED automotive headlamp lighting system based on a free-form surface. *Appl. Opt.* **2021**, *60*, 8910–8914. [CrossRef] [PubMed]
12. Wang, H.; Chen, Z.J.; Wu, H.; Ge, P. Free-form LED automotive front-illuminated optical lens design method. *Infrared Laser Eng.* **2014**, *43*, 1529–1534.
13. Hsieh, C.C.; Tsai, C.L.; Hung, C.C.; Lin, H.H.; Li, T.C. Design of Multi-module LED Headlamp of Vehicle Under Federal Motor Vehicle Safety Standard. *Sens. Mater.* **2023**, *35*, 2433–2488. [CrossRef]
14. Alfred, M.; Hossein, A. Folded light source: A technique for compact and efficient automotive lighting. *Opt. Eng.* **2023**, *62*, 095104.
15. *GB 25991-2010*; Automotive Headlamps with LED Light Sources and/or LED Modules. Standardization Administration of China (SAC): Beijing, China, 2022.
16. Oshima, R.; Masashige, S. Study of LED Spot Beam lighting for Headlights. In Proceedings of the 2019 IEEE International Conference on Consumer Electronics (ICCE), Las Vegas, NV, USA, 11–13 January 2019.
17. Tsai, C.Y. Design of free-form reflector for vehicle LED low-beam headlamp. *Opt. Commun.* **2016**, *372*, 1–13. [CrossRef]
18. Wu, H.; Zhang, X.M.; Ge, P. Design method of a light emitting diode front fog lamp based on a freeform reflector. *Opt. Laser Technol.* **2015**, *72*, 125–133. [CrossRef]
19. Wang, L.P.; Su, P.; Ma, J.; Huang, J. LED high-beam module design for automotive headlight. *AOPC 2021 Disp. Technol.* **2021**, *12063*, 20–27.
20. Mas, A.; Martín, I.; Patow, G. Heuristic driven inverse reflector design. *Comput. Graph.* **2018**, *77*, 1–15. [CrossRef]
21. Tan, M.Y.; Chen, J. Dynamical optimized lamps distribution based on genetic algorithm for visible light communications. In Proceedings of the Eleventh International Conference on Information Optics and Photonics, Xi'an, China, 20 December 2019; pp. 1016–1021.
22. Brunnstrom, K.; Andrew, J.S. Genetic algorithms for free-form surface matching. In Proceedings of the 13th International Conference on Pattern Recognition, Vienna, Austria, 25–29 August 1996; pp. 689–693.
23. Shewchuk, J.R. Delaunay refinement algorithms for triangular mesh generation. *Comput. Geom.* **2002**, *22*, 21–74. [CrossRef]
24. Rahbek, L.W.; Kirkegaard, P.H.; Alibrandi, U. Parametric grid mapping design tool for freeform surfaces using a genetic algorithm. *Proc. IASS Annu. Symp.* **2020**, *22*, 1–12.
25. Lu, S.T. Characterization of Fabric Moisture Absorption and Quick-Drying Based on Segmentation Genetic Algorithm. Master's Thesis, Zhongyuan University of Technology, Zhengzhou, China, 2023.
26. Li, X.L. Research on the Key Technology of Magnetorheological Polishing of Complex Curved Aluminum Mirror. Master's Thesis, National University of Defense Technology, Changsha, China, 2018.
27. Xu, C.; Peng, X.Q.; Dai, Y.F. Current status of ultra-precision manufacturing of complex curved aluminum reflectors. *Opto-Electron. Eng.* **2020**, *47*, 200147.
28. Xie, Y.J.; Mao, X.L.; Li, J.P.; Wang, F.; Wang, P.; Gao, R.; Li, X.; Ren, S.; Xu, Z.; Dong, R. Optical design and fabrication of an all-aluminum unobscured two-mirror freeform imaging telescope. *Appl. Opt.* **2020**, *59*, 833–840. [CrossRef] [PubMed]

photonics

MDPI

Article

Characteristics of Fused Silica Exit Surface Damage by Low-Temporal Coherence Light Irradiation

Chong Shan [1,2], Ping Han [2], Erxi Wang [1], Fujian Li [1], Xiaohui Zhao [1,*], Huamin Kou [2], Dapeng Jiang [2], Qinghui Wu [2], Xing Peng [3], Penghao Xu [4], Yafei Lian [5], Yuanan Zhao [5], Liangbi Su [2], Zhan Sui [2] and Yanqi Gao [1]

1 Shanghai Institute of Laser Plasma, China Academy of Engineering Physics, 1129 Chenjiashan Road, Shanghai 201800, China; shan_chong@163.com (C.S.); erxi_wang@163.com (E.W.); lifujian0824@163.com (F.L.); liufenggyq@siom.ac.cn (Y.G.)

2 State Key Laboratory of Functional Crystals and Devices, Shanghai Institute of Ceramics, Chinese Academy of Sciences, Shanghai 201899, China; hanping202504@163.com (P.H.); huaminkou@mail.sic.ac.cn (H.K.); dajiang@mail.sic.ac.cn (D.J.); wuqinghui@mail.sic.ac.cn (Q.W.); suliangbi@mail.sic.ac.cn (L.S.); lqling@vip.163.com (Z.S.)

3 College of Intelligent Science and Technology, National University of Defense Technology, Changsha 410073, China; pengxing22@nudt.edu.cn

4 Troops 91007, Chinese People's Liberation Army, China; zhenqing20072008@sina.com

5 Key Laboratory of High-Power Laser Materials, Shanghai Institute of Optics and Fine Mechanics, Chinese Academy of Sciences, Shanghai 201800, China; lianyafei@siom.ac.cn (Y.L.); yazhao@siom.ac.cn (Y.Z.)

* Correspondence: xhzhao_silp@163.com

Abstract: Laser-induced exit surface damage of fused silica is a key bottleneck for its application in high-power laser devices. As low-temporal coherence light (LTCL) has garnered increasing attention for high-power laser-driven inertial confinement fusion, understanding LTCL-induced exit surface damage of fused silica becomes crucial for improving the output power capability of LTCL devices. In this study, we characterized damage on the exit surface of fused silica under LTCL irradiation and investigated the physical mechanism of temporal coherence affecting the laser-induced damage threshold (LIDT). The relationship between defect information and temporal coherence was explored using a defect analysis model, and the defect damage process and response to each incident lasers were captured using time-resolved methods and artificially fabricated defects. We elucidate the physical mechanism behind the lower LIDT under LTCL irradiation compared to single longitudinal mode (SLM) pulse lasers. This study not only provides the boundary condition for safe fused silica operation in high-power LTCL devices but also offers deeper insight into the physical properties of LTCL.

Keywords: low-temporal coherence light; laser-induced damage threshold; time-resolved methods; artificially fabricated defects

1. Introduction

A positive fusion exergy output of 3.05 MJ at Livermore National Laboratory in the United States marks the successful "ignition" of laser inertial confinement [1,2]. The condition for achieving laser inertial confinement fusion (ICF) is very strict. The triple-frequency fluence irradiation on the terminal optical components has almost reached its safe operation limitation [3]. Fused silica has a wide range of applications in high-power laser devices due to its excellent optical, thermal, and mechanical properties [4–6]. Therefore, the issue of laser-induced fused silica damage is a bottleneck problem that restricts its application and even limits the laser output capability. In particular, laser-induced damage on the exit surface of fused silica is mainly caused by surface and sub-surface defects [7–9].

There is an electric field enhancement effect on the exit surface of fused silica during laser irradiation, resulting in a lower laser-induced damage threshold (LIDT) on the exit surface than on the input surface and bulk damage [8].

To mitigate the negative effects of laser plasma instability (LPI) and enable new ignition options, the low-temporal coherence light (LTCL) has garnered significant attention due to its characteristics, such as instantaneous broadband and random phase distribution [10–14]. Low-temporal coherence indicates a broadband pulse spectrum with a random spectral phase, also known as spectrally incoherent broadband pulses [15]. Such pulses, characterized by possession of both low-temporal coherence and instantaneous broadband properties, have the full spectral component in a time slice of the pulse coherence time scale [13]. These pulses could also be called low-temporal coherence light. In particular, recent experimental results demonstrate that LTCL can effectively suppress LPIs, such as stimulated Raman scattering, greatly enhancing ignition probability [16,17]. The LTCL has a broad bandwidth with a coherence time much shorter than the pulse duration, and its spectral phases are randomly and uniformly distributed, unlike those of chirped pulses. This matches the statistical properties of polarized thermal light. The LTCL can be viewed as an accumulation of many pulses, each with a duration close to the coherence time [11]. Nevertheless, laser-induced exit surface damage of fused silica upon LTCL irradiation poses a significant challenge for high-power LTCL device operation.

There has been extensive excellent research on the exit surface damage mechanisms of fused silica for conventional coherence light [18–25]. B.C. Stuart et al. modified the incident laser pulse width by changing the dispersive path length of a single-grating compressor. They determined the relationship between the LIDT of the exit surface and the incident pulse width, and established an analysis model to investigate multiphoton ionization effects [26,27]. Numerous studies also exist on the laser-induced damage characteristics for multiple longitudinal mode (MLM) pulse laser irradiation [28–31]. R. Diaz, et al. found that the laser-induced damage density on the surface of fused silica is higher for MLM pulse laser irradiation than for the single longitudinal mode (SLM) pulse laser at 1064 nm, and explained this by three-photon absorption by temporal spikes of the MLM pulse laser [29]. The LTCL has more complex spike structures than the MLM pulse laser, posing a greater threat to the fused silica surface's damage resistance and complicating the physical mechanism of defect damage. However, the physical mechanism of LTCL-induced exit surface damage of fused silica and the effect of temporal coherence on the LIDT have not been reported. Therefore, investigating the exit surface damage characteristics of fused silica by LTCL irradiation is a significant priority for its application in high-power LTCL devices.

In this study, we characterized the pattern of temporal coherence affecting the LIDT on the exit surface of fused silica. Utilizing the defect analysis model and the time-resolved method, the defect information corresponding to each bandwidth of LTCLs is obtained. Through artificially fabricated defects, we explored the mechanism of temporal coherence affecting the absorption of the defects and the LIDTs on the exit surface of fused silica. This study not only investigates the effect of temporal coherence on the properties of the LIDT on the exit surface of fused silica but also provides reliable defect precursor information for improving the LTCL-induced damage threshold of fused silica.

2. Experimental Details

Experimental Setup

In this study, a 40 mm × 40 mm × 3 mm fused silica wafer was used as the test sample. Both the input and exit surfaces were polished and acid washed to ensure the surface roughness and profile were maintained at 0.5 nm (RMS) and $\lambda/4$ (λ = 632.8 nm), respectively. The LIDT test setup for the exit surface of the fused silica sample is shown in Figure 1. The incident laser could produce either the SLM pulse laser or the LTCL with a wavelength center at 1053 nm. For the LTCL, the temporal coherence could be adjusted by controlling the spectral width of the incident laser, and the spectral full width at half-maximum (FWHMs) for the LIDT test were 3 nm, 8 nm, and 13 nm, respectively (Figure 2a). The central wavelengths of the SLM pulse laser, the LTCL (3 nm), the LTCL (8 nm), and the LTCL (13 nm) were 1053 nm, 1053 nm, 1062 nm, and 1058 nm, respectively. The difference in the central wavelength of each incident laser would affect the LIDT test results through differences in photon energy. However, based on the LIDT test result of fused silica described below, the difference in central wavelength has a lesser impact compared to the large differences in LIDT observed for each incident laser. The central wavelength effect on the LIDT for other optical materials would be analyzed in more detail in subsequent studies. The pulse duration of each incident laser was 3 ns (FWHM), as shown in Figure 2b.

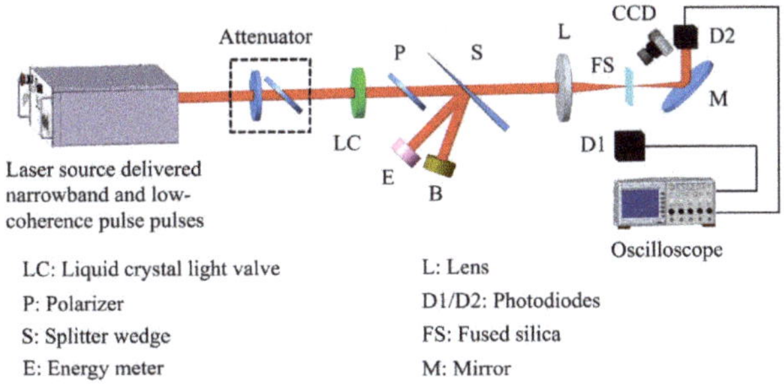

Figure 1. Schematic diagram of the LIDT test system.

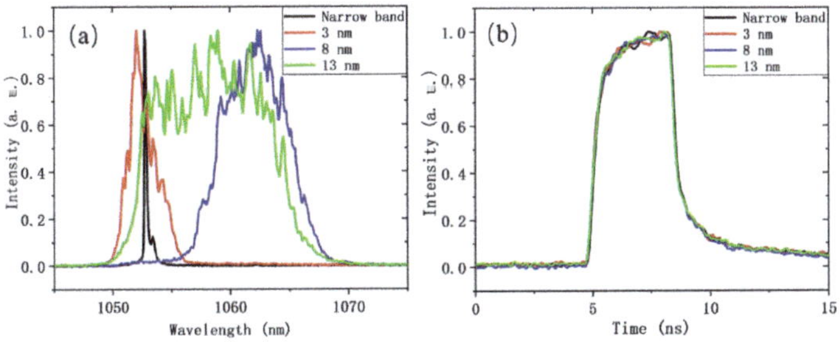

Figure 2. The (**a**) spectrum and (**b**) temporal waveforms of the incident pulses.

The incident energy for each incident laser is controlled by an attenuator consisting of a half-wave plate and a polarizer. The original near fields of each incident laser were slightly different. To ensure the accuracy of the experimental test, a combination of liquid crystal light value and polarizer was applied to regulate the near field of each incident laser, which made the beam area and energy distribution remain uniform. Through 10 times tests, the size of each incident laser was kept at (10 ± 0.2) mm at $1/e^2$, and the near-field energy distribution of each incident laser was fitted with a Gaussian function to ensure that this energy distribution was strictly maintained as a Gaussian distribution, which ensured the consistency of the near field of each incident laser.

The incident laser energy and beam qualities are monitored by an energy meter and a beam quality analyzer, respectively. A lens with a focal length of 4.5 m focuses the beam (the beam effective area at the focal point of 0.45 mm^2, and the incident laser fluence was equivalent to the incident laser energy divided by the effective area.), ensuring the Rayleigh length is longer than the sample thickness. Laser-induced surface damage of fused silica is recorded by a CCD camera on the side of the sample. Two InGaAs photodiodes, D1 and D2 (time resolution of 150 ps), connected to an oscilloscope sampling rate of 5 GHz, are placed on the side of the fused silica and in the direction of laser transmission for recording the damage process using the time-resolved method, and as the oscilloscope did not employ an interpolation algorithm, the overall temporal resolution of this time-resolved system was 200 ps. Invisible damage in fused silica was traditionally assessed by two methods: analyzing variations in the temporal waveform (recorded transmission signal by photodiodes) and the beam quality (recorded by a beam analyzer) during subsequent laser irradiation with low fluence after LIDT shots. If there was no variation in the temporal and spatial domains, it could be considered that no invisible damage had occurred. The LIDTs for each incident laser are tested using the 1-on-1 test method [32].

3. Experimental Results

The LIDT test results for each incident laser are shown in Table 1. It is found that the LIDTs on the exit surface of fused silica for the LTCLs are lower than those of the SLM pulse laser, and for the LTCL, the LIDTs decrease with increasing bandwidth. The corresponding damage mechanisms will be analyzed in the following section.

Table 1. The LIDTs of exit surface by each incident laser irradiation.

	SLM (J/cm^2)	LTCL-3 nm (J/cm^2)	LTCL-8 nm (J/cm^2)	LTCL-3 nm (J/cm^2)
LIDT	59.93	53.61	32.83	21.79

The damage morphologies of the fused silica exit surface by LTCL irradiation were recorded using a Scanning Electron Microscope (SEM: FEI- versa 3d, Hillsboro, OR, USA). To demonstrate the damage morphology clearly, the laser irradiation was carried out with a fluence (86.63 J/cm^2) much higher than the LIDT, as shown in Figure 3. It can be observed that the exit surface damage morphologies by LTCL irradiation are consistent with those of the conventional SLM pulsed laser [9], both of which consist of a melt central zone and a fractured periphery. The melt central zone is primarily caused by defect precursors on the fused silica surface and subsurface, which have a stronger absorption ability of incident laser energy compared to the fused silica substrate. As the defect precursor continuously absorbs laser energy, its temperature rises rapidly, and a melt central zone forms when the temperature reaches the melting point, presenting a relatively rough surface, as shown in Figure 3b. In the process of forming the melt central zone, temperature differences and corresponding thermal stresses are generated between the defect precursor and the surrounding fused silica substrate. These thermal stresses accumulate as the temperature of the defect precursor rises and ultimately remove material by mechanical stripping, presenting a smoother surface than the melt central zone, as shown in Figure 3c.

To obtain more defect damage information, we recorded the processes of laser-induced damage by the time-resolved method. Laser-induced surface damage is mainly divided into two damage phases: activation and expansion (as shown in Figure 4). The first phase is the incubation phase, mainly the process of defects on the surface or subsurface absorbing laser energy. When the absorption of laser energy exceeds the LIDT of the defect precursor and induced damage, the laser energy would be significantly absorbed and scattered by the defect damage, inducing a rapid decrease in the transmission signal. This is recorded by the photodiode (D_2) in the transmission direction (transmission signal) as the damage time (t_d) [24]. The second phase is the damage expansion process, which develops from the initial damage time (t_d) to the end of the pulse. Therefore, the total incident fluence (F_{tot}) could be written as follows:

$$F_{tot} = F_{inc} + F_{exp} = I_{peak}\int_{0}^{t_d} f(t)dt + I_{peak}\int_{t_d}^{+\infty} f(t)dt \quad (1)$$

where F_{tot} is the total fluence of incident laser, F_{inc} and F_{exp} represent the fluence absorbed by the defect precursors in the first process and the fluence required for the expansion of the damage in the second process, respectively. I_{peak} and $f(t)$ represent the incident peak intensity and a function of temporal waveform, respectively. We analyzed the defect precursor information of fused silica for each LTCLs by this method below.

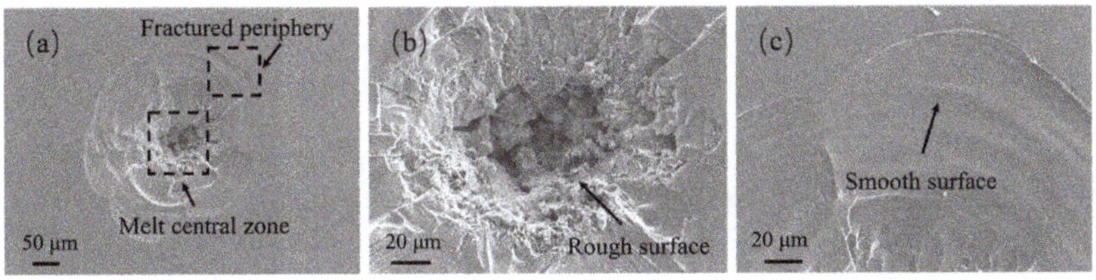

Figure 3. Damage morphology of fused silica exit surface recorded by SEM: (**a**) overall damage morphology of fused silica by the LTCL irradiation with 86.63 J/cm^2; (**b**) the melt central zone; (**c**) the fractured periphery.

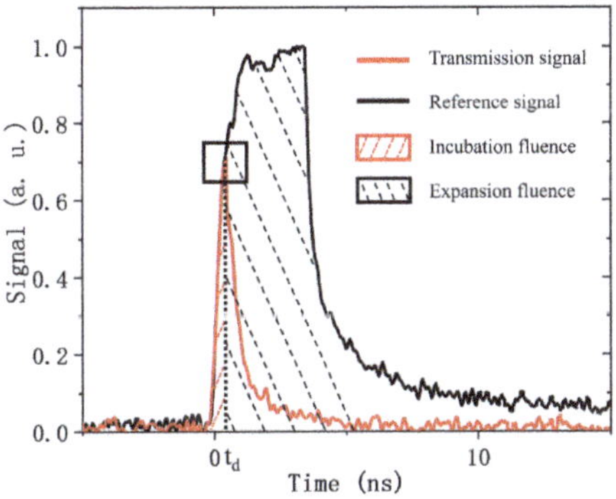

Figure 4. The results of transmission signal variation by time-resolved test method.

4. Discussion

4.1. Temporal Spike Structure

The reduction of LIDTs for the LTCLs is caused by a combination of several factors. Firstly, the LTCL possesses a more complex temporal spike structure than that of the MLM pulse laser, as shown in Figure 5, and these temporal spike structures would be one of the important factors affecting the LIDTs. A. V. Smith et al. concluded that the reduction in the LIDT for the MLM pulse laser is directly related to the maximum intensity during the temporal spike structure [28]. For the LTCLs, the distribution of the temporal spike intensities for each bandwidth satisfied the negative exponential distribution [14]. The highest intensity of the temporal spikes should be consistent for each LTCL, which should result in the same LIDTs for different bandwidths and should be 1/10 of the SLM pulse laser. However, the LIDT test results of the LTCLs during our experiments did not show a significant decrease like that, and they were not consistent for each bandwidth. This indicates that the highest intensity of the temporal spikes was not influenced by the LIDT directly, and there were other physical mechanisms during the laser-induced damage.

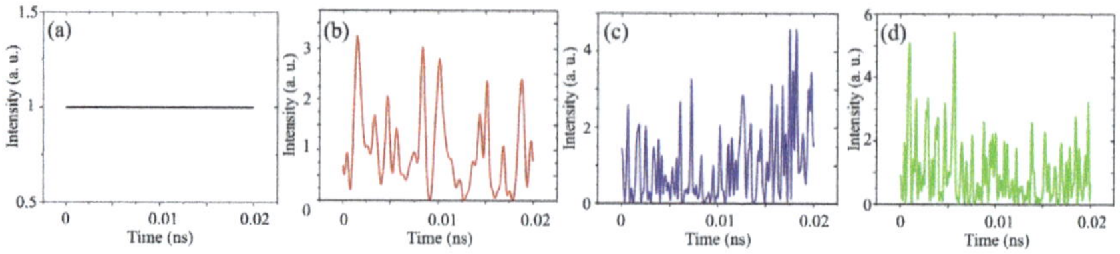

Figure 5. The schematic diagram of the temporal spike structure: (**a**) the SLM pulse laser; (**b**) 3 nm (LTCL); (**c**) 8 nm (LTCL); (**d**) 13 nm (LTCL).

Although the decrease in the LIDTs on the exit surface of fused silica is not directly induced by the highest intensity spikes for the LTCLs, the temporal spike structures also affect the LIDT test results by the multiphoton absorption and the accumulative effect. For the LTCLs, the spike structure in the temporal domain could be regarded as a series of pulse strings, and these multiple pulses present the cumulative effect, which would induce defects [33] and reduce the bandgap [34], which reduces the LIDTs of the fused silica [35]. The widths of the temporal spikes for the LTCLs are roughly 280 fs, 460 fs, and 1.25 ps, corresponding to the bandwidth of 13 nm, 8 nm, and 3 nm, respectively. (Since the experimental time-resolved setup was unable to capture the temporal spike widths of the LTCL with different bandwidths, these widths were obtained through theoretical calculations based on statistical properties of the LTCLs and the Fourier transform of their spectra.) This implies that the wider the bandwidth, the smaller the duration of the spikes, and the corresponding number of the spikes and the repetition frequency of the pulse train would be increased at the same duration of the incident pulse, which leads to a more significant decrease in the LIDTs of the fused silica [36]. It also verifies the previous speculation about why the MLM pulse laser decreased the LIDT of SiO_2 [37].

Meanwhile, due to the large bandgap of fused silica (approximately 9 eV) [38], multiphoton absorption with a fundamental frequency laser (1.17 eV) is improbable for the incident intensities involved in this paper. However, sub-bandgap photons might be absorbed by subsurface atomic-scale defects [39]. The strong photoluminescence under 3.5 eV excitation [39,40] is consistent with three-photon absorption of the fundamental frequency laser. It could be deduced that the LTCL, with temporal spike structures similar to the MLM pulse laser, would reduce the LIDT of fused silica through the multiphoton absorption effect. As mentioned above, the number of high-intensity temporal spikes increases with the bandwidth for the same pulse duration, promoting multiphoton absorption and further decreasing the LIDT, consistent with the relationship of the LIDT of LTCLs with different bandwidths. It was worth noting that the LIDT test results, as shown in Table 1, indicate that the difference between the LTCL (3 nm) and the SLM-pulsed laser is not significant. It is considered that there were relatively few temporal spike structures for the LTCL (3 nm) due to its large temporal spike durations; thus, nonlinear absorption effects, such as three-photon absorption, would not be effectively amplified. However, the LIDT of LTCL (8 nm) shows a significant decrease compared to LTCL (3 nm), suggesting that the multiphoton absorption effect has begun to dominate and affect the LIDT test results at this point.

4.2. Defect Precursor Information

Laser-induced surface damage of fused silica is primarily caused by defect precursors on the surface or sub-surface. Consequently, the different information of the defect precursors corresponding to the LTCLs and SLM pulse laser would also be a critical factor for influencing the LIDT test results. To study the effect of temporal coherence on defect precursors, a statistical model is assumed where the probability of a defect precursor absorbing higher fluence than its intrinsic damage threshold follows the Poisson law [41–44].

The damage probability related to the average number of defect precursors, N(F), can be expressed as follows:

$$P(F) = 1 - \exp[-N(F)] \tag{2}$$

where N(F) is the number of defect precursors in the beam spot with an intrinsic damage threshold below F:

$$N(F) = \int_0^F g(T)\cdot V_T(F)dT \tag{3}$$

The defect precursor distribution typically could be assumed to satisfy the Gaussian fluence distribution with a mean threshold fluence T_0, a standard deviation ΔT (full width at $1/e^2$), and defect precursor density [43,44]. The ensemble function g(T) could be expressed as follows:

$$\begin{cases} g(T) = \frac{2n_0}{\Delta T\sqrt{2\pi}} \exp\left[-\frac{1}{2}\left(\frac{T-T_0}{\frac{\Delta T}{2}}\right)^2\right] \\ \int_0^\infty g(T)dT = n_0 \end{cases} \tag{4}$$

The efficient volume $S_T(F)$ where the fluence F is higher than the intrinsic damage threshold of the defect precursor could be expressed as follows:

$$S_T(F) = S_{eff}\ln\left(\frac{F}{T}\right) \tag{5}$$

where the S_{eff} is the effective area of the incident laser beam. The correlation best-fit curves using the model with a coefficient of the determination R^2 of 0.92–0.95 are shown in Figure 6, and the defect precursors information on the exit surface of fused silica for each incident lasers are listed in Table 2. Parameters of this distribution (defect precursor densities (n_0), threshold mean value (T_0) and threshold standard deviation (ΔT) could be obtained clearly.

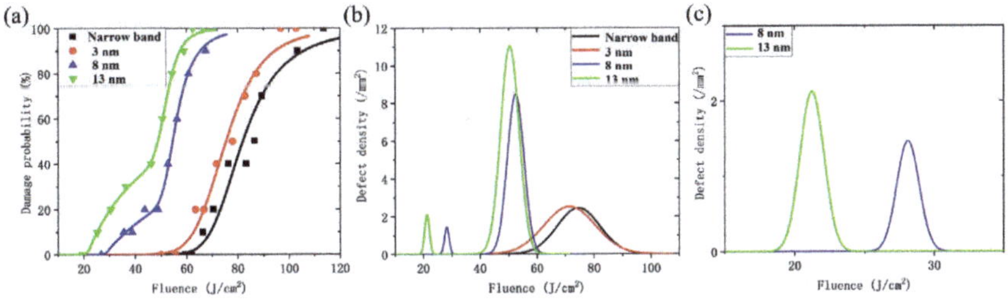

Figure 6. (**a**) Probability curves calculated with modeling data, dots denote experimental data, and solid lines denote fitted curves for each incident laser; (**b**) defect information from fitting the defect analysis model; (**c**) defect information of low damage probability for SLD pulse laser with 8 nm and 13 nm bandwidths.

Table 2. Defect information of fused silica surface under different incident laser irradiations.

Sequence	n_0 (mm^{-3})	T_0 (J/cm^2)	ΔT (J/cm^2)
Narrow band	46.26	74.58	14.99
SLD (3 nm)	57.06	71.36	17.96
SLD (8 nm)-a	2.97	28.10	1.61
SLD (8 nm)-b	62.59	52.30	5.88
SLD (13 nm)-a	4.49	21.19	1.68
SLD (13 nm)-b	89.80	50.24	6.48

It was found that only one kind of defect precursor was observed for both the SLM pulse laser and LTCL (3 nm). The defect density (n_0) of the LTCL (3 nm) was higher than that of the SLM pulse laser, the corresponding threshold mean value (T_0) for the LTCL (3 nm) was lower than that of the SLM pulse laser, and the threshold standard deviation (ΔT) of the LTCL (3 nm) had a larger distribution than that of the SLM pulse laser. This indicates that the LTCL (3 nm) corresponds to a wider variety of defect precursors (different sizes and depths from the surface) and a greater number of precursors on the fused silica surface than the SLM pulse laser, with a lower average threshold. For the LTCL (8 nm) and LTCL (13 nm), defect information at a high damage probability was similar to the LTCL (3 nm) and the SLM pulse laser, but a new type of defect appeared at a low damage probability. It was mainly caused by the fact that the temporal spike structure widths of the LTCL (8 nm) and LTCL (13 nm) were 460 fs and 280 fs, respectively, containing a much larger number of temporal spike structures with high peak intensity than the LTCL (3 nm) and SLM pulsed laser. This led to defect damage with strong absorption properties under low fluence irradiation, inducing a new type of defect damage information. The pattern of defect information aligned with LIDT test results, i.e., a wider bandwidth for the LTCL correlated with a higher defect density and lower average defect threshold.

To verify the accuracy of the analytical model for describing defect precursors, the time-resolved method was utilized to confirm the relationship between the average thresholds of the defect precursors corresponding to each incident laser. The incident laser fluence was applied between 80 and 120 J/cm^2 for the test irradiation of 20 points. The incubation fluence (F_{inc}) and expansion fluence (F_{exp}) were determined by recording the initial damage time (t_d) as described above. The experimental results are presented in Figure 7.

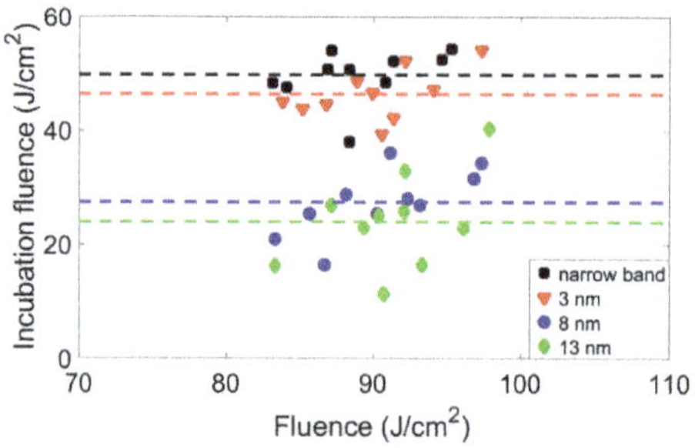

Figure 7. Incubation fluence as a function of the total fluence for each incident laser.

It was concluded that the average value of incubation fluence (F_{inc}) of the LTCL is lower than that of the SLM pulse laser, and the distribution of the incubation fluence (F_{inc}) is wider for the same range of total incident fluence. As mentioned above, the damage size is related to the expansion fluence (F_{exp}), while the incubation fluence (F_{inc}) is associated with the defects' properties. These differences are consistent with the relationship between the average defect damage threshold (T_0) difference obtained by utilizing the defect analysis model as described above, which verified the accuracy of the defect analysis model for characterizing the defect precursor information of the LTCLs.

4.3. Artificially Prepared Defect Precursor

Finally, it is considered that the LTCLs and SLM pulse laser have different absorptions for the same defect precursors, which also induced a decrease in the LIDT of the LTCL. To objectively represent the response of different incident lasers to the same defect precursors, artificial fabrication of the same defect precursors was performed for each incident laser. Due to the random distribution of defect precursor types and damage thresholds on the fused silica surface, it was difficult to find precursors with identical sizes and depths. Femtosecond lasers have been applied extensively due to their ultra-short and ultra-intense properties, such as in meta-surface preparation [45,46]. Here, pit defect precursors were prepared using a femtosecond laser to investigate the difference in absorption and damage threshold between the LTCLs and SLM pulse lasers.

The pit defects prepared by femtosecond laser were observed by SEM and FIB, as shown in Figure 8. Each pit is spaced 10 μm and 5 μm apart, respectively (as shown in Figure 8a). Their diameters and depths are 2.43 μm and 0.6 μm, respectively (as shown in Figure 8c). These artificially prepared pit defects have excellent consistency. It is worth noting that to amplify the difference in the absorption and response of each incident laser for the same defect precursors, and more easily compare the damage processes of the different lasers inducing the same defect damage, the initial defect size and depth prepared in this chapter were slightly larger than those introduced by conventional machining [39]. The simulation results for defect information (as shown in Table 2) indicated that LTCLs were more sensitive to defects with low damage thresholds. Therefore, it was considered that artificially prepared defects would be larger than typical natural precursors, corresponding to stronger absorption properties. This would amplify the LIDT difference between the LTCLs and the SLM pulsed laser.

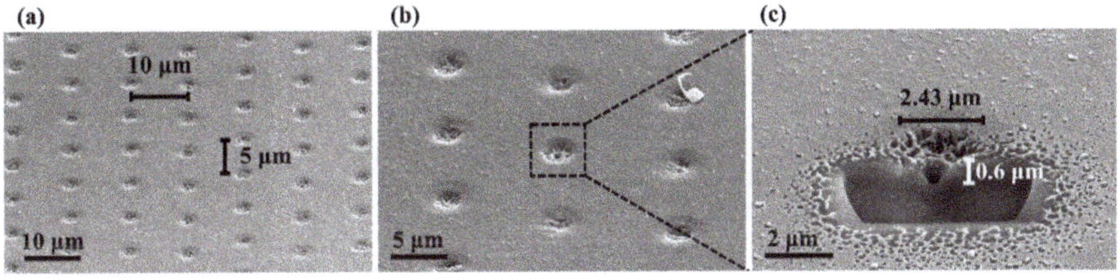

Figure 8. The initial defect morphologies by femtosecond laser preparation: (**a**) and (**b**) SEM; (**c**) FIB.

The pit defects were irradiated by each incident laser, and the defect damage morphology is shown in Figure 9. The pit damage morphologies consist of a melt central zone, a fractured periphery, and plasma ablation. The damage process and mechanism of the melt central zone and fractured periphery have been described above. For the plasma ablation area, it was mainly caused by the stronger absorption properties of the artificially prepared pit defects compared to the clean fused silica surface. Thus, the plasma generated during surface damage would expand outwards continuously, eventually forming the plasma ablation damage.

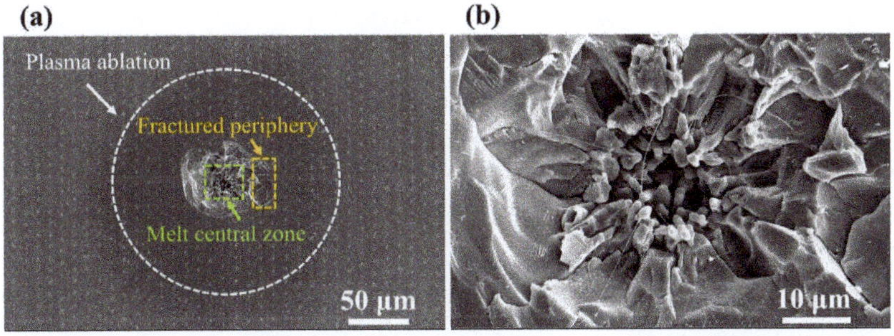

Figure 9. Artificially pit defect damage observed by SEM (**a**) the whole damage morphology of pit defects; (**b**) the melt explosion damage morphology in the central area of the beam spot.

The International LIDT Test Standard (ISO 21254) defines laser-induced damage in optical components as irreversible changes after laser irradiation, considered as laser-induced component damage. In this chapter, three damage morphologies—melt central zone, fractured periphery, and plasma ablation—were observed for artificially prepared defects upon laser irradiation, with corresponding damage processes and mechanisms described in the original manuscript. Therefore, using three material variations as criteria, the incident fluence for each laser was adjusted to determine the fluence at which pit defects did not produce a melt central zone, fractured periphery, and plasma ablation simultaneously, which was set as the LIDT of the pit defects. Test results are shown in Table 3. For the same defect precursors, the LIDTs for the LTCLs are lower than for the SLM pulse laser. For the LTCLs, a broader bandwidth corresponds to a more significant decrease in the LIDTs of pit defects. This indicates that LTCLs have a stronger reaction to the same defects than the SLM pulse laser, one of the main reasons for the difference in LIDTs between the LTCL and SLM pulse laser.

Table 3. The LIDTs of artificial defect precursor by each incident lasers irradiations.

	SLM (J/cm^2)	**LTCL-3 nm (J/cm^2)**	**LTCL-8 nm (J/cm^2)**	**LTCL-13 nm (J/cm^2)**
LIDT	4.07	3.28	2.63	1.37

5. Conclusions

In this paper, we systematically investigate the laser-induced damage characteristics on the exit surface of fused silica under LTCL irradiation. The test results suggest that the LIDTs of the LTCLs are lower than those of the SLM pulse laser, and for the LTCLs, the broader the bandwidth, the lower the LIDTs. Then, we analyzed the physical mechanism of the LIDT reduction for the LTCLs, concluding it is not directly caused by the maximum intensity of the temporal spike structure, but instead by multiphoton absorption and the cumulative effect of multiple pulses generated by the temporal spike structure. Analysis of defect density, average defect damage threshold, and defect threshold distribution also indicates larger differences in defect precursor information for LTCLs compared to SLM pulse lasers, contributing to decreased LIDTs. Finally, the different LIDTs of the artificially prepared pit defects precursor reveal that the LTCLs exhibit stronger absorption and response compared to the SLM pulse laser, also inducing a lower LIDT on the fused silica surface. This paper derives the characteristics and physical mechanism of the LTCL-induced fused silica surface damage and provides assistance in the investigation of the temporal coherence affecting the laser-induced damage mechanisms.

Author Contributions: Conceptualization, C.S. and P.H.; methodology, E.W.; software, F.L. and E.W.; validation, C.S., X.Z., H.K. and D.J.; formal analysis, X.P. and Q.W.; investigation, C.S. and Y.L.; resources, P.X., Y.Z., L.S., Z.S. and Y.G.; data curation, C.S. and P.H.; writing—original draft preparation, C.S.; writing—review and editing, X.Z. and F.L.; visualization, E.W. and F.L.; supervision, X.Z.; project administration, P.X. and Z.S.; funding acquisition, L.S., Y.G. and Z.S. All authors have read and agreed to the published version of the manuscript.

Funding: This research was supported by the National Natural Science Foundation of China (52450255).

Institutional Review Board Statement: Not applicable.

Informed Consent Statement: Not applicable.

Data Availability Statement: Data underlying the results presented in this paper are not publicly available at this time but may be obtained from the authors upon reasonable request.

Conflicts of Interest: The authors declare no conflicts of interest.

References

1. Clery, D. Explosion marks laser fusion breakthrough. *Science* **2022**, *378*, 1154–1155. [CrossRef] [PubMed]
2. Tollefson, J.; Gibney, E. Nuclear-Fusion Lab Achieves 'Ignition': What Does It Mean? *Nature* **2022**, *612*, 597–598. [CrossRef] [PubMed]
3. Manes, K.R.; Spaeth, M.L.; Adams, J.J.; Bowers, M.W.; Bude, J.D.; Carr, C.W.; Conder, A.D.; Cross, D.A.; Demos, S.G.; Di Nicola, J.M.G.; et al. Damage Mechanisms Avoided or Managed for NIF Large Optics. *Fusion Sci. Technol.* **2016**, *1*, 146–249. [CrossRef]
4. Miquel, J.-L.; Lion, C.; Vivini, P. The laser mega-Joule: LMJ & PETAL status and program overview. *J. Phys. Conf. Ser.* **2016**, *688*, 012067.
5. Zheng, W.; Wei, X.; Zhu, Q.; Jing, F.; Hu, D.; Yuan, X.; Dai, W.; Zhou, W.; Wang, F.; Xu, D.; et al. Laser performance upgrade for precise ICF experiment in SG-III laser facility. *Matter Radiat. Extrem.* **2017**, *2*, 243–255. [CrossRef]
6. Aikens, D.M.; Bissinger, H.D. Overview of small optics for the National Ignition Facility. In *Optical Manufacturing and Testing III*; SPIE: Bellingham, WA, USA, 1999; Volume 3782, pp. 476–487.
7. Camp, D.W.; Kozlowski, M.R.; Sheehan, L.M.; Nichols, M.A.; Dovik, M.; Raether, R.G.; Thomas, I.M. Subsurface damage and polishing compound affect the 355-nm laser damage threshold of fused silica surfaces. In *Laser-Induced Damage in Optical Materials*; SPIE: Bellingham, WA, USA, 1998; Volume 3244, pp. 356–364.
8. Tesar, A.A.; Brown, N.J.; Taylor, J.R.; Stolz, C.J. Subsurface polishing damage of fused silica: Nature and effect on laser damage of coated surfaces. In *Laser-Induced Damage in Optical Materials*; SPIE: Bellingham, WA, USA, 1991; Volume 1441, pp. 154–172.
9. Bude, J.; Miller, P.; Baxamusa, S.; Shen, N.; Laurence, T.; Steele, W.; Suratwala, T.; Wong, L.; Carr, W.; Cross, D.; et al. High fluence laser damage precursors and their mitigation in fused silica. *Opt. Express* **2014**, *22*, 5839–5851. [CrossRef]
10. Bates, J.W.; Myatt, J.F.; Shaw, J.G.; Follett, R.K.; Weaver, J.L.; Lehmberg, R.H.; Obenschain, S.P. Mitigation of cross-beam energy transfer in inertial-confinement-fusion plasmas with enhanced laser bandwidth. *Phys. Rev. E* **2018**, *97*, 061202. [CrossRef]
11. Cui, Y.; Gao, Y.; Rao, D.; Liu, D.; Li, F.; Ji, L.; Shi, H.; Liu, J.; Zhao, X.; Feng, W.; et al. High-energy low-temporal-coherence instantaneous broadband pulse system. *Opt. Lett.* **2019**, *44*, 2859. [CrossRef]
12. Gao, Y.-Q.; Ji, L.; Zhao, X.; Cui, Y.; Rao, D.; Feng, W.; Xia, L.; Liu, D.; Wang, T.; Shi, H.; et al. High-power, low-coherence laser driver facility. *Opt. Lett.* **2020**, *45*, 6839–6842. [CrossRef]
13. Ji, L.; Zhao, X.; Liu, D.; Gao, Y.; Cui, Y.; Rao, D.; Feng, W.; Li, F.; Shi, H.; Liu, J.; et al. High-efficiency second-harmonic generation of low-temporal-coherent light pulse. *Opt. Lett.* **2019**, *44*, 4359–4362. [CrossRef]
14. Zhao, X.; Ji, L.; Liu, D.; Gao, Y.; Rao, D.; Cui, Y.; Feng, W.; Li, F.; Shi, H.; Shan, C.; et al. Second-harmonic generation of temporally low-coherence light. *APL Photonics* **2020**, *5*, 091301. [CrossRef]
15. Dorrer, C.; Hill, E.M.; Zuegel, J.D. High-energy parametric amplification of spectrally incoherent broadband pulses. *Opt. Express* **2020**, *28*, 451–471. [CrossRef]
16. Lei, A.; Kang, N.; Zhao, Y.; Liu, H.; An, H.; Xiong, J.; Wang, R.; Xie, Z.; Tu, Y.; Xu, G.; et al. Reduction of Backward Scatterings at the Low-Coherence Kunwu Laser Facility. *Phy. Rev. Lett.* **2024**, *132*, 035102. [CrossRef]
17. Wang, P.P.; An, H.H.; Fang, Z.H.; Xiong, J.; Xie, Z.Y.; Wang, C.; He, Z.Y.; Jia, G.; Wang, R.R.; Zheng, S. Backward scattering of laser plasma interactions from hundreds-of-joules broadband laser on thick target. *Matter Radiat. Extrem.* **2024**, *9*, 015602. [CrossRef]
18. Crisp, M.D.; Boling, N.L.; Dube, G. Importance of Fresnel reflections in laser surface damage of transparent dielectrics. *Appl. Phys. Lett.* **1972**, *21*, 364–366. [CrossRef]
19. Laurence, T.A.; Bude, J.D.; Ly, S.; Shen, N.; Feit, M.D. Extracting the distribution of laser damage precursors on fused silica surfaces for 351 nm, 3 ns laser pulses at high fluences (20–150 J/cm^2). *Opt. Express* **2012**, *20*, 11561–11573. [CrossRef] [PubMed]
20. Sun, L.; Liu, H.; Huang, J.; Ye, X.; Xia, H.; Li, Q.; Jiang, X.; Wu, W.; Yang, L.; Zheng, W. Reaction ion etching process for improving laser damage resistance of fused silica optical surface. *Opt. Express* **2016**, *24*, 199–211. [CrossRef] [PubMed]
21. Lia, B.; Hou, C.; Tian, C.; Guo, J.; Xiang, X.; Jiang, X.; Wang, H.; Liao, W.; Yuan, X.; Jiang, X.; et al. Layer by layer exposure of subsurface defects and laser-induced damage mechanism of fused silica. *Appl. Surf. Sci* **2020**, *508*, 145186. [CrossRef]
22. Carr, C.W.; Trenholme, J.B.; Spaeth, M.L. Effect of temporal pulse shape on optical damage. *Appl. Phys. Lett.* **2007**, *90*, 041110. [CrossRef]
23. Palmier, S.; Gallais, L.; Commandre, M.; Cormont, P.; Courchinoux, R.; Lamaigne, L.; Rullier, J.L.; Legros, P. Optimization of a laser mitigation process in damaged fused silica. *Appl. Surf. Sci.* **2009**, *255*, 5532–5536. [CrossRef]
24. Xu, Y.; Emmert, L.A.; Rudolph, W. Spatio-TEmporally REsolved Optical Laser Induced Damage (STEREO LID) technique for material characterization. *Opt. Lett.* **2015**, *23*, 21607–21614. [CrossRef]
25. Miller, P.E.; Suratwala, T.I.; Bude, J.D.; Laurence, T.A.; Shen, N.; Steele, W.A.; Feit, M.D.; Menapace, J.A.; Wong, L.L. Laser damage precursors in fused silica. In *Laser-Induced Damage in Optical Materials*; SPIE: Bellingham, WA, USA, 2009; Volume 7504, pp. 318–331.
26. Stuart, B.C.; Feit, M.D.; Rubenchik, A.M.; Shore, B.W.; Perry, M.D. Laser-induced damage in dielectrics with nanosecond to subpicosecond pulses. *Phys. Rev. Lett.* **1995**, *74*, 2248. [CrossRef] [PubMed]

27. Stuart, B.C.; Feit, M.D.; Herman, S.; Rubenchik, A.M.; Shore, B.W.; Perry, M.D. Nanosecond-to-femtosecond laser-induced breakdown in dielectrics. *Phys. Rev. B* **1996**, *53*, 1749. [CrossRef] [PubMed]
28. Smith, A.V.; Do, B.T. Bulk and surface laser damage of silica by picosecond and nanosecond pulses at 1064 nm. *Appl. Opt.* **2008**, *47*, 4812–4832. [CrossRef]
29. Diaz, R.; Chambonneau, M.; Courchinoux, R.; Grua, P.; Luce, J.; Rullier, J.; Natoli, J.; Lamaignère, L. Influence of longitudinal mode beating on laser-induced damage in fused silica. *Opt. Lett.* **2014**, *39*, 674–677. [CrossRef] [PubMed]
30. Diaz, R.; Chambonneau, M.; Grua, P.; Rullier, J.-L.; Natoli, J.-Y.; Lamaignèrea, L. Influence of vacuum on nanosecond laser-induced surface damage morphology in fused silica at 1064 nm. *Appl. Surf. Sci.* **2016**, *362*, 290–296. [CrossRef]
31. Lamaignère, L.; Diaz, R.; Chambonneau, M.; Grua, P.; Natoli, J.-Y.; Rullier, J.-L. Correlation between laser-induced damage densities of fused silica and average incubation fluences at 1064 nm in the nanosecond regime. *J. Appl. Phys.* **2017**, *121*, 045306. [CrossRef]
32. *ISO 21254-1–21254-4*; Lasers and Laser-Related Equipment—Test Methods for Laser-Induced Damage Threshold Part 4: Inspection, Detection and Measurement. International Organization for Standardization: Geneva, Switzerland, 2011.
33. Mero, M.; Clapp, B.; Jasapara, J.; Rudolph, W.; Ristau, D.; Starke, K.; Krueger, J.; Martin, S.; Kautek, W. On the damage behavior of dielectric films when illuminated with multiple femtosecond laser pulses. *Opt. Eng.* **2005**, *44*, 051107. [CrossRef]
34. Gao, E.; Xie, B.; Xu, Z. Two-dimensional silica: Structural, mechanical properties, and strain-induced band gap tuning. *J. Appl. Phys.* **2016**, *119*, 014301. [CrossRef]
35. Natoli, J.Y.; Bertussi, B.; Commandre, M. Effect of multiple laser irradiations on silica at 1064 and 355 nm. *Opt. Lett.* **2005**, *30*, 1315–1317. [CrossRef]
36. Nguyen, D.N.; Emmert, L.A.; Patel, D.; Menoni, C.S.; Rudolph, W. Transient phenomena in the dielectric breakdown of HfO_2 optical films probed by ultrafast laser pulse pairs. *Appl. Phys. Lett.* **2010**, *97*, 191909. [CrossRef]
37. Chmel, A.E. Fatigue laser-induced damage in transparent materials. *Mat. Sci. Eng. B* **1997**, *49*, 175–190. [CrossRef]
38. Saito, K.; Ikushima, J. Absorption edge in silica glass. *Phys. Rev. B* **2000**, *62*, 8584–8587. [CrossRef]
39. Laurence, T.A.; Bude, J.D.; Shen, N.; Feldman, T.; Miller, P.E.; Steele, W.A.; Suratwala, T. Metallic-like photoluminescence and absorption in fused silica surface flaws. *Appl. Phys. Lett.* **2009**, *94*, 151114. [CrossRef]
40. Fournier, J.; Néauport, J.; Grua, P.; Fargin, E.; Jubera, V.; Talaga, D.; Jouannigot, S. Green luminescence in silica glass: A possible indicator of subsurface fracture. *Appl. Phys. Lett.* **2012**, *100*, 114103. [CrossRef]
41. Natoli, J.Y.; Gallais, L.; Akhouayri, H.; Amra, C. Laser-induced damage of materials in bulk, thin-film, and liquid forms. *Appl. Optic.* **2002**, *41*, 3156–3166. [CrossRef] [PubMed]
42. Krol, H.; Gallais, L.; Grezes-Besset, C.; Natoli, J.Y.; Commandre, M. Investigation of nanoprecursors threshold distribution in laser-damage testing. *Opt. Commun.* **2005**, *256*, 184–189. [CrossRef]
43. Bataviciute, G.; Grigas, P.; Smalakys, L.; Melninkaitis, A. Revision of laser-induced damage threshold evaluation from damage probability data. *Rev. Sci. Instrum.* **2013**, *84*, 045108. [CrossRef]
44. Liao, Z.M.; Roussell, R.; Adams, J.J.; Runkel, M.; Frenk, W.T.; Luken, J.; Carr, C.W. Defect population variability in deuterated potassium di-hydrogen phosphate crystals. *Opt. Mater. Express* **2012**, *2*, 1612–1623. [CrossRef]
45. Makarov, S.V.; Tsypkin, A.N.; Voytova, T.A.; Milichko, V.A.; Mukhin, I.S.; Yulin, A.V.; Putilin, S.E.; Baranov, M.A.; Krasnok, A.E.; Morozov, I.A.; et al. Self-adjusted all-dielectric metasurfaces for deep ultraviolet femtosecond pulse generation. *Nanoscale* **2016**, *8*, 17809–17814. [CrossRef]
46. Shcherbakov, M.R.; Liu, S.; Zubyuk, V.V.; Vaskin, A.; Vabishchevich, P.P.; Keeler, G.; Pertsch, T.; Dolgova, T.V.; Staude, I.; Brener, I.; et al. Ultrafast all-optical tuning of direct-gap semiconductor metasurfaces. *Nat. Commun.* **2017**, *8*, 17. [CrossRef] [PubMed]

photonics

MDPI

Article

SparsePose–NeRF: Robust Reconstruction Under Limited Observations and Uncalibrated Poses

Kun Fang [1], Qinghui Zhang [1,*], Chenxia Wan [1], Pengtao Lv [1] and Cheng Yuan [2]

[1] College of Information Science and Engineering, Henan University of Technology, Zhengzhou 450001, China; 202191025@stu.haut.edu.cn (K.F.); wanchenxia@haut.edu.cn (C.W.); pengtaolv@163.com (P.L.)

[2] Henan Center for Fair Competition Review Affairs, Zhengzhou 450008, China; hnipsc@163.com

* Correspondence: zqh131@haut.edu.cn

Abstract

Neural Radiance Fields (NeRF) reconstruction faces significant challenges under non-ideal conditions, such as sparse viewpoints or missing camera pose information. Existing approaches frequently assume accurate camera poses and validate their effectiveness on standard datasets, which restricts their applicability in real-world scenarios. To tackle the challenge of sparse viewpoints and the inability of Structure-from-Motion (SfM) to accurately estimate camera poses, we propose a novel approach. Our method replaces SfM with the MASt3R-SfM algorithm to robustly compute camera poses and generate dense point clouds, which serve as depth–space constraints for NeRF reconstruction, mitigating geometric information loss caused by limited viewpoints. Additionally, we introduce a high-frequency annealing encoding strategy to prevent network overfitting and employ a depth loss function leveraging Pearson correlation coefficients to extract low-frequency information from images. Experimental results demonstrate that our approach achieves high-quality NeRF reconstruction under conditions of sparse viewpoints and missing camera poses while being better suited for real-world applications. Its effectiveness has been validated on the Real Forward-Facing dataset and in real-world scenarios.

Keywords: NeRF; sparse-view reconstruction; MASt3R-SfM; uncalibrated camera poses; real-world applications; uncalibrated poses

1. Introduction

NeRF, a neural network-based technique, has achieved remarkable success in 3D scene reconstruction and novel view synthesis in recent years. By learning from multi-view image data, NeRF [1] can generate high-quality 3D scene representations and demonstrates significant potential in applications such as virtual reality, augmented reality, and film production. However, despite its superior performance under ideal conditions, NeRF faces numerous challenges when applied to real-world scenarios, particularly in 3D reconstruction under non-ideal conditions. Non-ideal conditions generally refer to significant deficiencies in input data, such as poor lighting, sparse viewpoints, inaccurate camera poses, incomplete image data, or noisy inputs. These issues are prevalent in practical applications and pose substantial challenges to the reconstruction accuracy and stability of NeRF [2–18].

Our work focuses on two specific conditions: sparse viewpoints and inaccurate camera poses. Sparse viewpoints limit the amount of training information available, making it challenging to capture the complete geometric structure and details of the scene. This limitation often results in blurred or distorted novel view images generated by NeRF. Furthermore, inaccurate camera poses can directly impact the quality of scene reconstruction

https://doi.org/10.3390/photonics12100962

and rendering, as NeRF training relies on precise camera poses (typically image extrinsics) to correctly estimate spatial relationships within the scene. Inaccurate poses may lead to stretching, distortion, or misalignment of objects, significantly compromising the quality of reconstruction results. In real-world applications, we observe that sparse viewpoints and the absence or inaccuracy of camera poses often occur simultaneously. This phenomenon substantially increases the complexity and difficulty of 3D reconstruction. Moreover, we find that existing NeRF improvement methods fail to address this issue effectively. Current NeRF improvement methods, whether targeting sparse viewpoints or camera pose optimization, are typically based on ideal assumptions, such as dense viewpoint coverage and precise camera poses. It should be noted that these studies primarily validate their methods on standard datasets, where dozens of views are used to compute accurate camera poses with COLMAP [19], which are then stored for direct use during reconstruction. These studies seem to overlook a critical issue: when only 3–5 views are available, camera poses estimated using SfM algorithms often contain significant noise. This is primarily because traditional SfM methods rely on limited matching points under sparse input viewpoints, inevitably leading to unstable camera pose estimation and significant deviations or even failures in NeRF reconstruction.

To address these challenges, this paper proposes a novel NeRF reconstruction method suitable for sparse viewpoints and conditions without camera poses. We introduce the MASt3R-SfM [20] algorithm as a replacement for the traditional SfM pipeline, enabling robust camera pose estimation from sparse inputs and generating dense point clouds to constrain NeRF's depth–space reconstruction and compensate for geometric information loss due to limited viewpoints. Additionally, we design a high-frequency annealing encoding strategy that progressively enhances the network's ability to learn high-frequency information, preventing overfitting and local optima during training. Finally, we propose a depth loss function based on Pearson correlation coefficients, combined with a ray-sampling strategy in low-frequency regions to further improve the network's learning capability under limited viewpoints. The proposed method was experimentally validated on the Real Forward-Facing dataset and real-world scenarios. Results demonstrate that this method achieves high-quality 3D reconstruction under non-ideal conditions, such as sparse viewpoints and absent camera poses, while better aligning with practical application requirements.

The main contributions of this paper are as follows:

- A novel NeRF reconstruction framework for sparse viewpoints and conditions without camera poses, effectively addressing geometric information loss and training instability in real-world scenarios.
- A high-frequency annealing encoding strategy designed to progressively enhance the network's ability to learn high-frequency information, preventing overfitting.
- The introduction of the MASt3R-SfM algorithm for robust camera pose estimation from sparse inputs and dense point cloud generation, serving as depth–space constraints for NeRF.
- A depth loss function based on Pearson correlation coefficients, combined with a ray-sampling strategy in low-frequency regions to explore low-frequency information in limited viewpoints.

2. Related Work

To tackle the challenges posed by sparse viewpoints and inaccurate or missing camera poses, researchers have proposed various enhancements to NeRF to reduce its reliance on precise input data and camera pose information. Some of these approaches integrate self-supervised camera pose optimization, allowing NeRF to reconstruct scenes without depending on accurate pose data. BARF [21] introduces the traditional Bundle Adjustment (BA) framework into NeRF training, enabling joint optimization of camera poses and scene parameters to iteratively refine initial pose estimates. This novel joint optimization framework effectively resolves the challenges of NeRF training under conditions of unknown or inaccurate camera poses by employing dynamic frequency encoding and a coarse-to-fine optimization approach. However, the performance of this coarse-to-fine strategy is highly contingent on the complexity of the task scene and the specific parameter configurations for frequency scheduling. NeRF-- [22] (NeRF minus minus) incorporates camera poses as learnable parameters during training, addressing the scene reconstruction problem under unknown camera conditions by leveraging scene consistency constraints and optimization. However, while NeRF-- enables camera pose learning, it suffers from slow training convergence, sensitivity to initial pose estimates, performance degradation under extremely sparse inputs, and limited robustness to dynamic scenes. NoPe–NeRF [23] is a pose-free NeRF training method that integrates monocular depth priors with multi-view geometric constraints. Using an innovative joint optimization framework, it effectively addresses the challenges of scene reconstruction and novel view rendering under extensive camera motion, providing novel insights for the application of NeRF in scenarios without pose information. NoPe–NeRF typically relies on 20–30 input views to achieve high-quality novel view rendering and scene reconstruction. Although it excels in sparse scenarios, its performance is constrained under extremely sparse input conditions.

Overall, these advancements in NeRF reconstruction without accurate camera poses still depend on multiple input images. However, in real-world applications, if a sufficient number of images is available, traditional SfM algorithms can be employed to achieve relatively accurate camera pose estimations. Consequently, NeRF-based approaches that do not require camera poses lose their practical value under such circumstances. In practical applications, we often have access to only a limited number of images, making it challenging for traditional SfM algorithms to accurately estimate camera poses. This necessitates the development of algorithms capable of accurately estimating camera positions and performing precise 3D reconstruction and novel view synthesis under sparse-view conditions.

Several recent advancements in NeRF-based methods have addressed some of these challenges under sparse input settings. SparseNeRF [24] introduces a new approach that extracts depth ordering and continuity priors from coarse depth maps, effectively mitigating reconstruction degradation under sparse viewpoints. However, this method has limited capacity for handling unobserved occluded regions, and the decline in the quality of coarse depth maps may affect the model's reconstruction performance. RegNeRF [25] enhances NeRF's interpolation capability for unobserved regions under sparse viewpoints by introducing several regularization techniques, such as view consistency and spatial consistency, thus reducing the reliance on complete viewpoint coverage. However, this method has limited ability to complete details in unobserved regions, which may result in blurry predictions. FreeNeRF [26] provides a simple baseline that requires no complex modifications, significantly improving NeRF performance under sparse viewpoints with just two lines of code changes: frequency regularization and occlusion regularization. It offers novel insights for future neural rendering research, particularly regarding the crucial role of frequency control in low-data environments. IBRNet [27] introduces a general and efficient multi-view rendering framework that combines dynamic feature extraction of IBR with NeRF's volumetric rendering to enable novel view generation for complex scenes. This method does not require scene-specific optimization and exhibits strong generalization ability. However, when input views are extremely sparse, rendering quality may degrade, particularly in complex geometric scenes. DS-NeRF [28] introduces a depth supervision loss to enhance NeRF's geometric reconstruction ability under sparse viewpoints. By utilizing sparse point clouds generated by SfM at no additional cost, it significantly improves both reconstruction quality and training efficiency. DDP-NeRF [29] significantly enhances NeRF's geometric reconstruction and rendering quality under sparse viewpoints by leveraging dense depth priors and uncertainty estimation. However, this method relies on sparse point clouds generated by SfM, and the quality of the input depth may affect the completion performance when the depth is inaccurate. Moreover, depth completion methods may need to be extended for dynamic or non-rigid scenes. DietNeRF [30] significantly enhances viewpoint synthesis performance under sparse viewpoints by incorporating semantic consistency supervision. It utilizes CLIP's [31] powerful visual encoding capabilities to achieve reasonable completion of unobserved regions while preserving geometric consistency, marking a significant breakthrough for sparse-viewpoint NeRF applications. However, while the semantic consistency loss converges quickly under sparse viewpoints, it struggles to capture high-frequency texture details. Furthermore, the performance of this method depends on the quality of the pre-trained visual encoder, which may limit its effectiveness in small or specific scenes. PixelNeRF [32] achieves high-quality novel view synthesis under sparse input conditions by leveraging image features and learning scene priors, offering significant advantages in reducing training time and enhancing generalization ability. However, this method may face challenges in maintaining geometric and appearance consistency in scenes with complex backgrounds or significant occlusions. MVSNeRF [33] innovatively combines multi-view stereo geometry reasoning and neural radiance field reconstruction, significantly enhancing reconstruction quality and generalization ability under sparse viewpoints, providing a practical solution for efficient novel view synthesis. The aforementioned studies have optimized and improved the NeRF algorithm under sparse input conditions. However, it is important to acknowledge a critical reality: in existing research exploring the application of NeRF under conditions of no camera pose and sparse viewpoints, notable limitations still persist. On one hand, no-pose NeRF methods rely on a large number of input views to compensate for the lack of camera pose information, but capturing such a large number of images in real-world scenarios is often impractical. On the other hand, sparse-view NeRF methods achieve good reconstruction results with fewer input views, but they typically assume accurate camera poses, which are often challenging to satisfy in real-world applications, particularly in scenarios with limited viewpoints or low image quality.

To address these challenges, this paper proposes a new method designed to tackle both no-pose and sparse-view conditions simultaneously. Our method enables high-quality novel view synthesis and geometric reconstruction with only a few input views and without the need for accurate camera poses. Furthermore, we further validate the effectiveness of the method on the Real Forward-Facing dataset and real-world captured scenes, showcasing its robustness and generalization ability under sparse-view and no-pose conditions.

3. Method

This section provides a detailed explanation of the proposed sparse-view NeRF implementation method. As illustrated in Figure 1, the method consists of three main components.

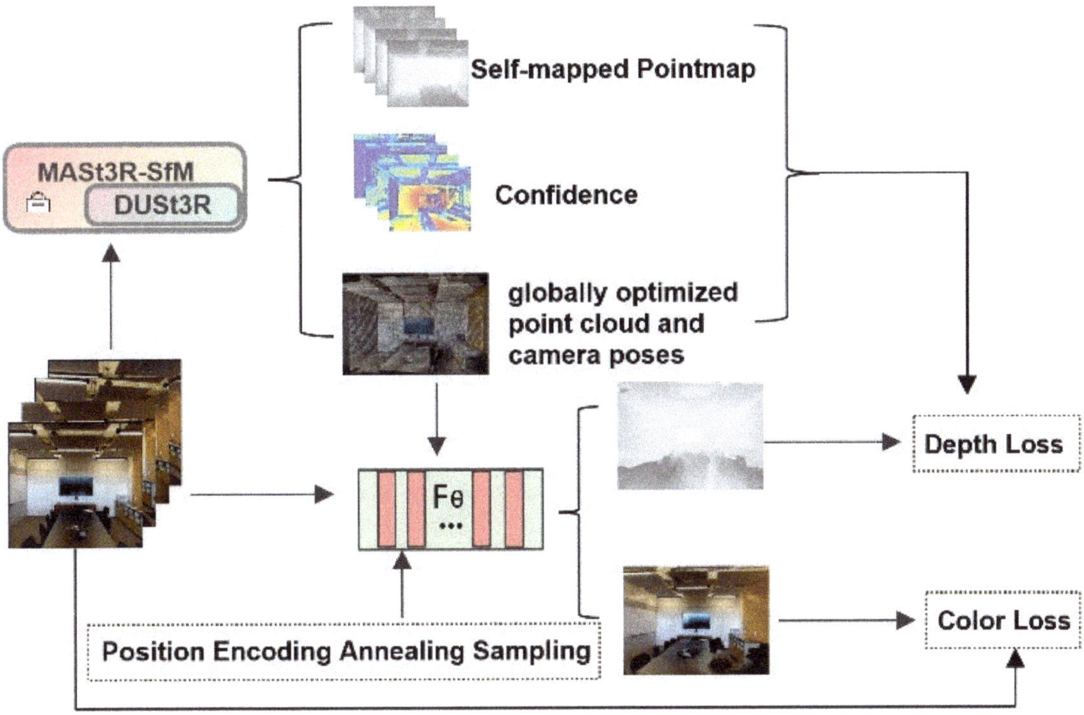

Figure 1. Overall Workflow Diagram of SparsePose–NeRF. Our method first utilizes the MASt3R-SfM frontend to robustly estimate camera poses from sparse input views while simultaneously generating an initial self-mapped point map and a final globally optimized point cloud. This geometric prior information is used to construct a two-stage depth loss. Concurrently, the NeRF network (FΘ) is trained using a position encoding annealing sampling strategy. Finally, the network is optimized by combining the depth loss and the traditional color loss to achieve high-quality reconstruction.

- Pose Estimation and Map Generation

The first component involves pose estimation based on the MASt3R-SfM algorithm and the generation of auxiliary maps, including an initial point cloud map, a normalized point map, and a confidence map.

- High-Frequency Annealing Encoding Strategy

The second component introduces a high-frequency annealing strategy for positional encoding in the NeRF network, ensuring that high-frequency overfitting does not occur during the early stages of training.

- Depth-Supervised Training Regularization

The third component focuses on regularizing the NeRF network training process to improve 3D reconstruction and novel view synthesis under sparse input conditions. To achieve this, two depth-related loss functions—denoted as L_U and L_M—are proposed.

3.1. The Principles and Characteristics of MASt3R-SfM and Its Role Within the Network

The MASt3R-SfM [20] network is essentially an extension of the MASt3R network's downstream tasks. It leverages the encoder and decoder of the pre-trained MASt3R [34] model to construct a more robust SfM algorithm. MASt3R is an enhanced version of the DUSt3R [35] network, better suited for high-precision matching and relative pose estimation tasks. DUSt3R's primary focus is on 3D reconstruction, which simultaneously predicts point maps and camera calibration parameters. Specifically, DUSt3R achieves 3D reconstruction by predicting dense 3D point maps, with matching relationships as byproducts. This approach relies on point regression for 3D reconstruction but is susceptible to noise, particularly in low-texture regions or under large baseline variations, leading to reduced matching accuracy. DUSt3R introduces a matching head to directly learn local feature descriptors instead of relying solely on 3D point map regression. This enhances feature matching robustness, especially in scenarios with significant viewpoint changes or repetitive patterns. Owing to its superior performance, MASt3R-SfM outperforms existing methods across multiple datasets and metrics, particularly excelling in sparse scenes and unordered image collections. Traditional SFM methods rely on keypoint matching to construct point trajectories across multiple images. However, MASt3R is inherently a pairwise image matcher. If traditional SFM methods are applied, the complexity grows quadratically with the number of images N. To address this issue, a minimal yet sufficient subset is constructed and input into MASt3R, where the subset structure forms a scene graph G. The graph G = (V,E) consists of vertices I ∈ V (representing images) and edges e = (n,m) ∈ E (representing undirected connections between potentially overlapping images In and I_m). This graph ensures all images are connected directly or indirectly. To construct an appropriate connected graph, MASt3R's encoder is used to predict the co-visibility score s ∈ (0,1) between two images, In and I_m. The detailed construction process is referenced in MASt3R-SfM. During runtime, the algorithm applies the MASt3R inference process to each edge e, generating the corresponding point maps and sparse pixel matches M_{nm}. In summary, the process represents the forward inference process of MASt3R, where the input is an image pair (I_n, I_m) and the output is the sparse feature matching M_{nm} between the two images. This approach also yields different point maps, such as X_{nn}, X_{nm}, X_{mn}, and X_{mm}, where X_{nn} and X_{mm} represent self-mapped point maps (i.e., 3D point maps generated within the coordinate system of a single image). These self-mapped point maps encode the 3D coordinates of pixels derived solely from intra-image information, independent of inter-image geometric relationships. The next step involves constructing canonical point maps through global fusion of estimated 3D scene points across images. This fusion eliminates viewpoint discrepancies and registration errors while ensuring global consistency, ultimately achieving a more accurate scene representation via weighted averaging. The canonical point map formulation is defined as follows:

$$\tilde{X}_n(i,j) = \frac{\sum_{e\in\varepsilon_n} C_{n,e}(i,j) X_{n,e}(i,j)}{\sum_{e\in\varepsilon_n} C_{n,e}(i,j)} \quad (1)$$

where

$\tilde{X}_n(i,j)$ is the canonical 3D point for pixel (i, j) of image In;

ε_n is the set of all edges in the scene graph connected to image In;

$X_{n,e}(i,j)$ is the estimated 3D point for pixel (i, j) of image In derived from the image pair associated with edge e;

$C_{n,e}(i,j)$ is the corresponding confidence value for that 3D point estimation.

For each pixel (i, j), the estimated values are averaged using confidence-based weighting. Through this formulation, a canonical point map is generated, integrating information from all associated image pairs. Additionally, the constrained point map is constructed based on the canonical point maps. The constrained point map rigorously enforces alignment of estimated 3D points (i.e., points generated from images) with a pinhole projection model. This constraint ensures that the point cloud adheres to the camera's geometric characteristics, thereby establishing a foundation for subsequent optimization. The constrained point map at pixel (i, j) is defined as follows:

$$\chi_n(i,j) = \frac{1}{\sigma_n} P_n^{-1} K_n^{-1} Z_n(i,j)[i,j,1]^T \quad (2)$$

where

$\chi_n(i,j)$ is the constrained 3D point in the world coordinate system for pixel (i, j);
K_n is the camera intrinsic matrix for image In;
$Z_n(i,j)$ is the depth value at pixel (i, j);
σ_n is a per-camera scale factor;
$[i,j,1]^T$ is the homogeneous coordinate of the pixel.

This equation ensures that the 3D points in the point cloud strictly conform to the pinhole projection model. The optimization objective is the minimization of the reprojection error. After normalization, the next step involves performing coarse alignment, which primarily aims to globally optimize and align the initial 3D point cloud to a unified world coordinate system. The optimization objective of coarse alignment is to minimize the loss function as follows:

$$\sum_{(m,n)\in\varepsilon} \sum_{c\in M^{n,m}} q_c\left(\rho\left(y_n^c - \pi_n\left(\chi_m^c\right)\right) + \rho\left(y_m^c - \pi_m\left(\chi_n^c\right)\right)\right) \quad (3)$$

where

q_c is the confidence of a given match c;
y_n^c and y_m^c are the 2D pixel coordinates of the match in images In and Im, respectively;
χ_m^c and χ_n^c are the corresponding 3D points;
π_n is the projection function for camera n;
ρ is a robust error function.

The above provides a concise summary of the MASt3R-SfM network's workflow and core optimization steps. Subsequently, our framework utilizes the self-mapped point clouds generated by the network, combined with globally optimized point clouds and camera poses, to achieve sparse-view NeRF reconstruction.

3.2. The Implementation Approach of Our Network and the Regularization Optimization Process

In NeRF, positional encoding is a crucial operation that maps the input 3D coordinates into a high-dimensional space, enabling the network to better capture high-frequency details. Conventional fully connected neural networks inherently prioritize low-frequency signal modeling, whereas NeRF explicitly targets high-frequency detail reconstruction (e.g., edges and textures) in 3D scenes. Through positional encoding, the input coordinates are transformed into multi-frequency signal representations, thereby facilitating MLPs' learning of high-frequency content. The positional encoding process utilizes a series of sine and cosine functions to project the input signals. For a 3D input point x = (x, y, z), the positional encoding γ(x) is defined as follows:

$$\gamma(x) = \left(x, sin\left(2^0\pi x\right), cos\left(2^0\pi x\right) \ldots sin\left(2^{L-1}\pi x\right), cos\left(2^{L-1}\pi x\right)\right) \quad (4)$$

where

$\gamma(x)$ is the positional encoding function for an input coordinate x;
x is a 3D inputpoint (x, y, z);
L is the number of frequency levels, controlling the amount of high-frequency components.

Here, L represents the number of frequency levels, which controls the quantity of high-frequency components in positional encoding. Under sparse-view conditions, the network excessively focuses on high-frequency information (e.g., surface textures or local geometric variations), causing the MLP to overfit these details. Due to insufficient low-frequency sampling, the scene's global geometry may become distorted, compromising the network's ability to model object shapes and structural coherence. This limitation further manifests as artifacts in lighting and material properties. Smooth lighting transitions and material consistency are poorly reconstructed, often resulting in localized brightness spikes or color discontinuities. The root cause lies in inappropriate extrapolation of high-frequency patterns to unobserved regions, inducing spurious textures in empty spaces. Consequently, reconstructed surfaces exhibit unnatural complexity while lacking low-frequency smoothness. To address these challenges, we introduce a frequency-annealed positional encoding strategy during NeRF training, defined as follows:

$$L_{max} = L \times \frac{3}{5} \times \left(1 - cos\left(\pi \frac{t}{T}\right)\right) \tag{5}$$

where

L_{max} is the maximum frequency level available to the network at the current training stage;
L is the total number of frequency levels in the encoding;
t is the current training step;
T is the total number of training steps per epoch.

Frequency encoding plays a crucial role in determining the performance of the Neural Radiance Fields during training. In the formula, T denotes the total number of training steps per epoch, L represents the number of frequency levels in the encoding, $\mathrm{L_{max}}$ indicates the maximum frequency level accessible to the network at the current stage, and t refers to the current training step. In most NeRF networks, L is typically capped at a maximum value of 10. Under this setting, it is generally assumed that when $\mathrm{L_{max}} \geq 6$, the network begins to capture high-frequency information, which is essential for reconstructing fine details in the scene. The annealing schedule gradually activates the network's ability to sample high-frequency information as the training process reaches its halfway point. Each step t corresponds to an update of the network's weights, achieved by computing gradients using a batch of data and updating the parameters accordingly. This approach forces the network to focus on limited low-frequency information during the early stages of training, compensating for the deficiencies caused by sparse viewpoints. Merely limiting the network's capacity to sample high-frequency information does not address the persistent challenge that low-frequency information remains constrained below a fixed threshold. This limitation makes it challenging for the network to learn additional low-frequency features effectively. Beyond sampling and learning low-frequency information in the color domain, we aim to enable the network to acquire low-frequency information about the scene in the depth domain. During the initial stages of network training, we propose introducing depth supervision by comparing the depth differences between adjacent pixels within predefined patches. This approach focuses on capturing depth loss at a patch-level granularity, rather than directly comparing depth values at the pixel level. There are two primary reasons for this approach:

- It injects additional low-frequency geometric priors to constrain training and mitigate sparse-view limitations.
- Pixel-level depth comparisons introduce high-frequency noise from estimation errors, destabilizing early training phases.

In practice, the Sobel operator computes horizontal (Gx) and vertical (Gy) gradients, which are combined to obtain the gradient magnitude Gmag for each pixel. To clearly illustrate the processing procedure, we visually illustrate the processing of high- and low-frequency information in images in Figure 2. We employed distinct color annotations to differentiate high-frequency and low-frequency regions in the results. Regions where the gradient fell below the defined threshold were designated as low-frequency zones, indicated in light red, whereas regions with gradients exceeding the threshold were labeled as high-frequency zones, depicted in dark green. In the early stages of network training, we implemented an annealing strategy for NeRF positional encoding, which limited the network's ability to learn high-frequency features. As a result, during the initial phase of training, avoiding the selection of pixels in high-frequency regions for sampling rays proved to be both resource-efficient and beneficial for faster network convergence. Our approach involves continuously sampling multiple patches from each low-frequency region, with each patch sized at 32 × 32 pixels, encompassing a total of 1024 pixels per patch. To ensure comprehensive coverage, we maintain a 30% overlap between sampled regions. If a given region contains fewer than 1024 pixels in total, it is skipped entirely. Unlike the Vanilla NeRF strategy, which randomly shuffles all training set pixels and selects them randomly during the early stages of training, our method ensures that all sampled pixels are contained within the same batch.

The purpose of this approach is to compare the depth map rendered by NeRF during training with the initial depth map derived from the self-mapped point cloud of each image and to construct a Pearson correlation coefficient loss. The rationale for utilizing the initial depth map instead of the depth information from the final normalized point cloud lies primarily in the pairwise-image depth estimation properties: these maps reflect matching relationships between image pairs and exhibit feature-aligned consistency. In contrast, the normalized point cloud represents a weighted average of multiple estimations, smoothing out depth details. The initial depth map provides globally coherent priors that facilitate low-frequency acquisition in NeRF-based 3D reconstruction. However, its scale ambiguity precludes similarity evaluation via L2 loss with mean-based comparisons. To address this, we employ the Pearson correlation coefficient to measure patch-wise depth similarity between NeRF-rendered values and MASt3R-SfM self-mapped depths. During early training stages, we extend the Vanilla NeRF framework by augmenting its color–space loss L_C with a depth–space Pearson loss L_U.

$$L_C = \frac{1}{N}\sum_{i=1}^{N}||\hat{C}_i - C_i||^2 \tag{6}$$

$$L_U = \frac{1}{N}\sum_{i=1}^{N}1 - PCC_{patch}(d_N, d_M) \tag{7}$$

$$PCC(d_N, d_M) = \frac{E(d_N d_M) - E(d_N)E(d_M)}{\sqrt{E(d_N^2) - E(d_N)^2}\sqrt{E(d_M^2) - E(d_M)^2}} \tag{8}$$

where

L_C is the color–space loss;
L_U is the depth–space Pearson loss;
PCC is the Pearson correlation coefficient function;
$\hat{C}_i$ is the predicted color for sample i and C_i is the ground truth color;
d_N is the NeRF-rendered depth map for a patch;
d_M is the corresponding depth map from the MASt3R-SfM self-mapped point cloud;
E(·) denotes the expectation (mean) operator.

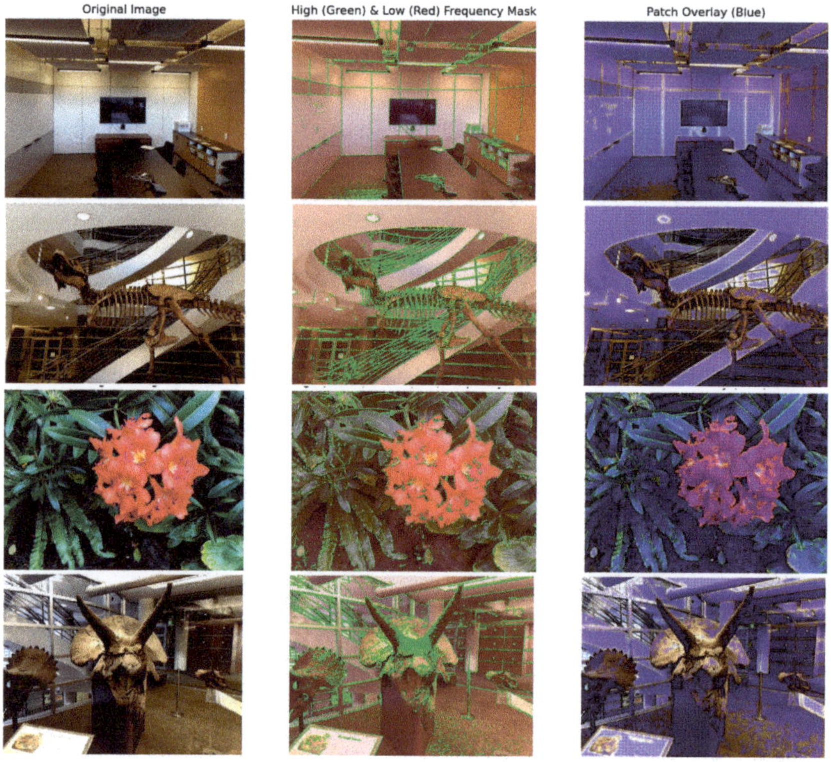

Figure 2. In the early stages of network training, we employed the Sobel operator to compute horizontal and vertical gradients, and the gradients from both directions were synthesized to obtain the gradient magnitude of each pixel. To visually illustrate this processing, we applied color-coded annotations to highlight high-frequency and low-frequency regions in the results. Specifically, regions with gradients below a predefined threshold were marked as low-frequency areas (light red), while regions with gradients exceeding the threshold were identified as high-frequency areas (dark green). Our sampling strategy involved continuously extracting multiple patches from each low-frequency region, with each patch sized 32 × 32 pixels (containing 1024 pixels). Additionally, we ensured 30% overlap between sampled regions. If the total number of pixels in a given region was fewer than 1024, that region was excluded from the sampling process.

In this loss function, N represents the total number of samples used for the calculation, and it denotes the Pearson correlation coefficient between d_N and d_M within the i-th patch. By minimizing this function, the correlation between the predicted depth d_N and the target depth d_M is pushed closer to 1, ensuring that both exhibit similar variation trends. Since the calculation of the Pearson correlation coefficient (PCC) is inherently normalized, this loss function is insensitive to global scale changes in depth values. By computing PCC within a patch, the function captures local correlations, which helps address issues arising from non-uniform depth variations across the scene. By maximizing local correlation, the method ensures correct depth variation trends within the scene, leveraging trend consistency to optimize the network in low-texture or low-confidence regions. As the number of training steps increases and reaches half of the total training steps, the annealing function gradually releases the network's capacity for high-frequency sampling. At this stage, the network begins to focus on high-frequency details in the scene. We introduced two significant changes to the network during this phase: First, we modified the ray sampling strategy within each batch. Initially, sampling was patch-based, where patches were randomly selected from each viewpoint. However, this approach contributed limitedly to the model's understanding of the global geometry of the scene. To address this, we adopted a random view sampling with rays strategy in the second half of the training. This adjustment improved the network's global geometric comprehension of the scene and prevented over-reliance on data from specific viewpoints. However, consistently using this sampling strategy might slow down the learning progress for details in certain viewpoints. To mitigate this, we incorporated importance sampling in regions of interest to increase sampling density in high-frequency areas of the scene, facilitating the capture of textures and edge details. Additionally, we introduced hybrid sampling, where a portion of rays were sampled uniformly across the entire scene, while others were concentrated in high-gradient regions, edges, or other important areas. By dynamically alternating among these three sampling strategies as training progressed, we significantly enhanced the training performance of NeRF while reducing unnecessary computational overhead. The second improvement involves an adjustment to the depth constraint by incorporating global-scale depth information. This constraint enables NeRF to align depth values learned from different viewpoints, effectively reducing errors and enhancing overall reconstruction capability. Previously, depth loss calculations were performed on individual depth maps without standardized scales. While MASt3R-SfM provided an initial global alignment through its self-mapped point cloud outputs in earlier stages, the precision was limited. In later stages, MASt3R-SfM further refined the alignment by conducting global fine optimization on top of coarse alignment, significantly improving the accuracy of camera poses and scene geometry, resulting in normalized point cloud information. To ensure alignment between NeRF's rendering results and the final normalized point cloud in terms of scale for loss computation, we adopted the following approach: At the start of network training, the sampling distances for near and far were carefully set. This allowed NeRF's depth information to align with the scale of the point cloud by matching the nearest and farthest points, thereby naturally aligning the scale of NeRF with the optimized point cloud output by MASt3R-SfM.

$$near = min(Z_{po}) - \xi \tag{9}$$

$$far = max(Z_{po}) + \xi \tag{10}$$

where

near and *far* are the near and far rendering bounds for NeRF;
Z_{po} represents the depth values of the globally optimized point cloud from MASt3R-SfM.

Here, ξ represents a small redundancy value introduced to prevent insufficient depth range caused by errors or occlusions. Subsequently, in the aligned depth–space, the difference between the rendered depth $d_{N,j}$ along each ray and the corresponding point cloud back-projected depth d_{Pj} is computed to formulate the loss function, which is defined by the following equation:

$$L_M = \frac{1}{N}\sum_{j=1}^{N}\left(\frac{(d_{N,j} - d_{P,j})^2}{c_j^2}\right) \tag{11}$$

where

L_M is the global-scale depth loss function;
N is the number of sampled rays in a batch;
$d_{N,j}$ is the depth rendered by NeRF along ray j;
$d_{P,j}$ is the corresponding depth from the back-projected globally optimized point cloud.

In the formula, the value c_j^2 in the denominator represents the square of the confidence level of the corresponding point cloud. This value reflects the uncertainty in the depth prediction of the pixel. Based on this uncertainty, a normalization operation is applied to the depth difference. This implies that when the predicted depth value exhibits high uncertainty, the influence of the error term is attenuated. Conversely, when the prediction uncertainty is low, the model imposes stricter penalties on large prediction errors. This design allows the model to relax error constraints in regions where predictions are inherently challenging. Finally, the total loss function for the NeRF network is defined as follows:

$$L_{ALL} = L_C + (1-\lambda)L_U + \lambda L_M \tag{12}$$

where

L_{ALL} is the total loss function for the NeRF network;

L_C, L_U, and L_M are the color loss, local Pearson depth loss, and global depth loss, respectively.

Where λ is a hyperparameter. At training step $t \leq T/2$, $\lambda = 0$, and at step $t > T/2$, it takes a value between 0 and 1. Based on the comprehensive analysis above, we have comprehensively annotated and visualized both the detailed execution workflow of SparsePose–NeRF and its corresponding loss functions at each step in Figure 3.

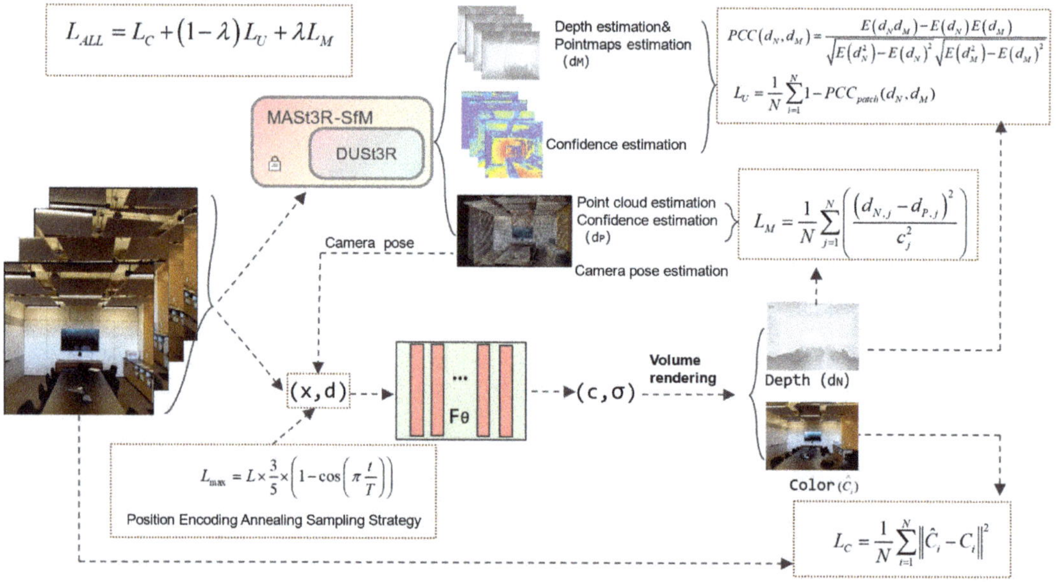

Figure 3. The input unconstrained images are first processed by the MASt3R-SfM network to compute camera poses, self-mapped point maps, confidence maps, and normalized point maps. The estimated camera poses are then passed to the NeRF network for ray sampling, where a position-encoded high-frequency quenching strategy is applied to prevent the network from overfitting to high-frequency information. During rendering, the depth maps generated by the NeRF network are compared with the precomputed depth values to construct the depth regularization loss functions L_M and L_U. Additionally, the RGB values rendered by the NeRF network are compared with the input images to formulate the color–space regularization loss function. The red, green, and blue lines denote the estimated camera poses generated by our model.

4. Experiments and Results

In the experimental section, we conducted three groups of experiments to evaluate the network's performance under different conditions. Specifically, the first set of experiments assessed the accuracy of pose estimation in complex environmental conditions. The second and third sets of experiments evaluated the network's ability to reconstruct dataset scenarios and real-world scenes under the dual adverse conditions of unknown camera poses and sparse viewpoints, as well as its capability to synthesize novel views.

All experiments were conducted on a workstation equipped with a 13th Gen Intel(R) Core(TM) i7-13700KF CPU and an NVIDIA GeForce RTX 4090 GPU. Our method follows a per-scene optimization paradigm. For each scene, we train our SparsePose–NeRF model for 200,000 iterations, a number consistent with related state-of-the-art works such as DietNeRF and FreeNeRF to ensure full convergence. The entire pipeline for a typical 3-view scene is highly efficient. The front-end pose and geometry estimation using MASt3R-SfM is completed in approximately 4 s. We also note that MASt3R-SfM reduces the computational complexity from quadratic to quasi-linear by constructing a sparse scene graph, as detailed in its original paper. The subsequent back-end NeRF optimization takes about 4.5 h. Once trained, rendering a new 800×800 resolution view takes approximately 5–10 s. This optimization time is competitive and notably faster than standard NeRF baselines, which can require over 10 h to converge on sparse data. The near and far rendering bounds were determined from the point cloud provided by the front end, with the redundancy margin ξ (Equations (9) and (10)) set to 5% of the scene's depth range. For the hyperparameter λ in our loss function (Equation (12)), which weights the global, confidence-based depth loss L_M, it is activated in the second half of training and linearly increases from 0 to its final value of 0.1. We found this setting to be robust across all scenes and datasets, requiring no per-scene tuning.

4.1. Validation of Camera Pose Estimation Accuracy

To validate the accuracy of the network in estimating camera poses, we conducted experiments on the Tanks and Temples [36] dataset, the CO3Dv2 [37] dataset, and the RealEstate10K [38] dataset. The Tanks and Temples dataset features a wide range of complex indoor and outdoor scenes, encompassing various lighting conditions, texture complexities, and scene scales. The design of the scenes in this dataset considers real-world complexities, such as non-uniform lighting, low-texture regions, and intricate geometric structures. These challenging characteristics make it an ideal benchmark for evaluating the robustness of algorithms. To quantitatively assess the accuracy of camera pose estimation, we compared our network with the COLMAP algorithm. It is worth noting that the COLMAP algorithm computes camera poses using all the images in each dataset, and its results are treated as ground truth for comparison purposes.

To calculate the Absolute Trajectory Error (ATE) [39], we align two camera trajectories using the Procrustes alignment method. This involves three key steps: decentering, normalization, and rotational alignment of both trajectories. After alignment, the translational error between the two poses is computed. We also evaluate the Relative Rotation Accuracy (RRA@5) and Relative Translation Accuracy (RTA@5) to assess the precision of relative rotations and translations between camera pairs. RRA@5 is a commonly used metric for high-precision rotation error evaluation. Considering that COLMAP performs remarkably well in high-overlap and high-texture scenes where it achieves highly accurate rotation estimations, we adopted a rigorous evaluation criterion. In addition, we report the registration success rate (Reg), which measures the proportion of cameras with successfully estimated poses. This metric is an essential indicator of an algorithm's robustness and applicability across various scenarios. In the Tanks and Temples dataset, we selected eight scenes and performed resampling and subdivision of the original dataset into subsets containing 25, 50, and 100 frames. Our algorithm was run under these three conditions, and the corresponding camera pose estimation metrics were reported, as summarized in Table 1.

Table 1. Quantitative experimental results of pose estimation on the Tanks and Temples dataset.

View Number	Scene	RRA@5	RTA@5	ATE	Reg
25 views	Ballroom	25.6	23.7	0.0616	100
	Barn	100	99.6	0.0006	100
	Church	45.8	50.8	0.0698	100
	Family	78.2	95.2	0.0043	100
	Francis	75.3	76.4	0.0176	100
	Horse	65.1	74.5	0.0053	100
	Ignatius	100	100	0.0002	100
	Museum	12.3	11.2	0.0969	100
50 views	Ballroom	24.0	19.2	0.0678	100
	Barn	100	99.9	0.0005	100
	Church	47.6	49.4	0.0707	100
	Family	81.2	96.7	0.0028	100
	Francis	45.2	60.3	0.0201	100
	Horse	65.9	75.9	0.0037	100
	Ignatius	100	100	0.0001	100
	Museum	20.3	22.6	0.0786	100
100 views	Ballroom	28.0	27.3	0.0377	100
	Barn	98.0	97.8	0.0020	100
	Church	63.3	63.4	0.0352	100
	Family	81.3	96.8	0.0019	100
	Francis	72.8	88.3	0.0027	100
	Horse	68.0	75.9	0.0026	100
	Ignatius	100	100	0.0001	100
	Museum	22.0	23.6	0.0502	100

The experimental results comprehensively validate the efficiency and robustness of our method in multi-view 3D reconstruction tasks across various view numbers (25, 50, and 100 views) and diverse scenes (e.g., Ballroom, Barn, and Church). As the number of views increases, the absolute trajectory error (ATE) significantly decreases, and both reconstruction accuracy (RRA@5) and time efficiency (RTA@5) consistently improve. For example, in geometrically rich scenes (e.g., Barn and Ignatius), our method exhibits near-perfect reconstruction results under all view configurations, with extremely low ATE values and RRA@5 and RTA@5 approaching 100%. Furthermore, even in complex scenes with sparse texture and geometry (e.g., Museum), our method maintains a 100% registration rate (Reg), demonstrating its robustness in handling global keypoints. In scenes with intricate structures and more dynamic changes (e.g., Family and Church), our method demonstrates superior capability in capturing finer details, and as the number of views increases, it significantly enhances accuracy and consistency, further narrowing the gap with baseline methods.

This performance indicates that our method not only excels in traditional static scenes but also adapts well to scenes with complex geometries and dynamic changes. Overall, these experimental results fully confirm the comprehensive advantages of our method in terms of accuracy, efficiency, robustness, and scene adaptability. Whether in geometrically rich, simple scenes or challenging scenarios with sparse texture or dynamic changes, our method consistently exhibits high stability and consistency, providing strong support for improving accuracy and adapting to complex scenes in 3D reconstruction tasks.

In the qualitative experiments, we utilized COLMAP to generate high-quality 3D reconstructions and camera pose estimations as ground truth. To evaluate the accuracy of our network's camera pose estimation, we compared the camera trajectories generated by our method against those produced by the state-of-the-art pose-free NeRF algorithm, NoPe–NeRF, across eight selected scenarios. During the experiments, we applied the Procrustes alignment method to align and visualize the camera trajectories, as illustrated in Figure 4. From the visualized results, it can be observed that our proposed method performs well on complex datasets such as Tanks and Temples. For instance, in most scenes (e.g., Ballroom, Barn, Francis, and Museum), the trajectories reconstructed by MASt3R-SfM are highly consistent with those of COLMAP, demonstrating superior accuracy and robustness. This indicates that our method can reliably recover 3D trajectories in these scenarios. In contrast, NoPe–NeRF exhibits significant deviations or instability in certain scenes (e.g., Horse and Church), highlighting its limitations in handling complex or structurally unique environments, possibly due to constraints in its model capacity or algorithmic design. Moreover, in scenes with intricate structures and variations (e.g., Family and Church), MASt3R-SfM outperforms NoPe–NeRF by capturing finer details and maintaining alignment with COLMAP's trajectories. Overall, these trajectory visualizations clearly illustrate the advantages of our algorithm in terms of camera pose estimation accuracy, stability, and adaptability to diverse scenes. However, the results also reveal that NoPe–NeRF requires further optimization to handle challenging scenarios effectively.

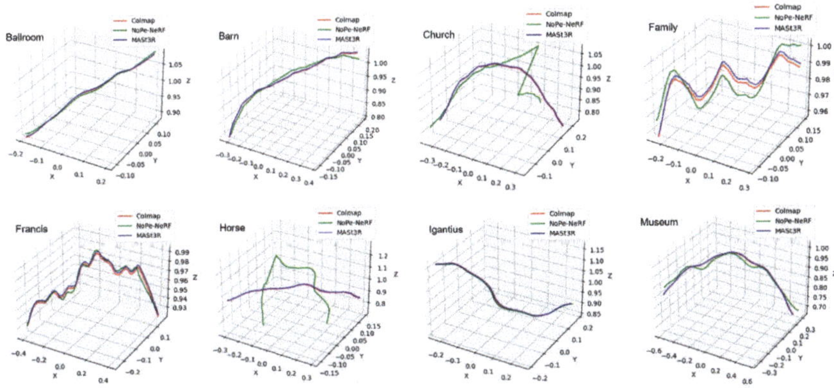

Figure 4. Trajectory prediction experiments were conducted on eight scenes from the Tanks and Temples dataset. The Procrustes alignment method was applied to align and visualize the camera trajectories, with the trajectories scaled to ensure their full extent could be displayed within the visualization. In the visualized plots, the camera poses estimated by COLMAP served as the ground truth. Comparisons were made between the camera trajectories predicted by the proposed method and those generated by the NoPe–NeRF algorithm.

To evaluate the accuracy of our network's camera pose estimation under sparse viewpoints, we conducted experiments on the CO3Dv2 and RealEstate10K datasets, both of which provide ground truth camera parameters. On the RealEstate10K dataset, we reported the mAA(30) metric, while on the CO3Dv2 dataset, we included RRA@15, RTA@15, and mAA(30). This choice is justified because RealEstate10K is primarily designed for novel view synthesis (NVS), focusing on the quality of generated views rather than precise pose estimation. The dataset typically involves views captured across large-scale scenes, which may lack highly accurate 3D pose annotations. The mAA(30) metric, being directly based on the matching quality of synthesized views, better reflects overall image-level performance compared to RRA and RTA, which are more reliant on precise camera poses. Furthermore, we set the threshold parameters for RRA and RTA to 15° because a rotational error within this range is generally acceptable in various computer vision tasks such as multi-view geometry, SLAM, and visual localization. For tasks like scene understanding or novel view synthesis, visual consistency is often maintained even with rotational errors within 15°. Additionally, practical applications such as indoor navigation and scene modeling do not demand extremely fine-grained rotational accuracy, making 15° a reasonable threshold to evaluate robustness in common scenarios. During the experiments, we selected 10 scenes from the test set of the RealEstate10K dataset. For each scene, three subsets were created by randomly sampling 3, 5, and 10 consecutive images, respectively. Similarly, from the CO3Dv2 dataset, we selected five categories: hotdog, motorcycle, bicycle, laptop, and toy bus. For each category, two shooting scenes were chosen, and subsets of varying sizes were constructed following the same approach used for the RealEstate10K dataset. Finally, the average parameter values were calculated across all scenes, and the results are presented in Table 2.

Table 2. Experiments were conducted on the RealEstate10K and CO3Dv2 datasets to evaluate the accuracy of camera pose estimation.

Frames	RealEstate10K		CO3Dv2	
	mAA(30)	RRA@15	RTA@15	mAA(30)
3	83.6	92.6	90.3	84.2
5	82.1	95.3	92.5	87.6
10	86.7	96.3	94.8	87.3

From the experimental results, it can be observed that our network achieves highly accurate camera pose estimation even under conditions of extreme sparsity in image data. This effectively addresses the challenge faced by NeRF networks in obtaining precise camera poses when operating with sparse viewpoints.

4.2. Comparative Experiments on LLFF Dataset with Sparse-View NeRF Models

We conducted comparative experiments on the LLFF [40] dataset with other sparse-view NeRF models. The experiments focused primarily on scene reconstruction and novel view synthesis using three input views. Both quantitative and qualitative analyses were performed to evaluate the results. Previous improvements in sparse-view NeRFs often evaluated their models by directly sampling a fixed number of images from datasets and utilizing the camera poses provided within those datasets for NeRF reconstruction. However, an overlooked fact is that these camera poses are estimated using the COLMAP algorithm by processing dozens of images of the same scene collectively. Under such conditions, the errors in camera pose estimation are minimal.

Accurate camera poses are crucial for NeRF reconstruction, as NeRF fundamentally relies on optimizing scene density and radiance fields. This optimization requires known camera poses to establish the origin and direction of rays, correlating 3D points in space with corresponding image pixels. In real-world applications, obtaining accurate camera poses under sparse-views is highly challenging. COLMAP typically requires at least five images to generate stable sparse point clouds. Moreover, it assumes sufficient overlap in the field of view and robust feature matching between images. When the number of images is limited or the overlap and viewing angle differences are minimal, the number of matched points decreases significantly. This results in incomplete scene reconstruction and inaccurate camera pose estimation. Through our experimental validation, we observed that existing methods for sparse-view reconstruction perform exceptionally well when provided with more than six input views. Therefore, our experiments focus on scenarios with three input views, where current NeRF-based methods for sparse-view improvements generally underperform. Even when these NeRF models directly utilize accurate camera poses from datasets—thus avoiding the challenges of inaccurate camera poses in real-world scenarios—their reconstruction quality and novel view synthesis remain suboptimal. This limitation is primarily due to insufficient scene information, which prevents the network from capturing adequate low-frequency information. By improving the NeRF network architecture, our proposed model effectively addresses these challenges and achieves superior performance. We compared four different methods for achieving NeRF-based 3D reconstruction under sparse view inputs. Due to experimental constraints, we used only three input views.

Directly using COLMAP to compute camera parameters from these three views often failed or resulted in significant camera pose errors. Furthermore, there were no ground truth images available for the new view synthesis experiments under these conditions. To address these challenges and facilitate experiments with three input views, we adopted a strategy of directly reading standard camera pose parameters from the dataset for the input to the comparison networks. Specifically, we selected three images from each of the four scenes in the Real Forward-Facing dataset as training inputs for NeRF and one additional image from the remaining images as ground truth for evaluating novel view synthesis performance. However, during the experiments with our network, we did not use the camera poses provided by the dataset; we consistently utilized our MASt3R-SfM frontend network to estimate camera poses. While our estimated poses exhibited minor deviations from those computed by COLMAP in the dataset, these differences had negligible effects on quantitative analysis. In qualitative experiments, the slight shifts in viewing angles were insignificant. This demonstrates that our method's pose estimation is a viable replacement for the traditional SfM algorithm in the NeRF pipeline.

The final results, as shown in Figure 5, highlight the performance differences across methods. For the "room" scene, PixelNeRF exhibited poor performance, with blurred boundaries on walls and ceilings and significant geometric distortions in objects on the desk and floor. DSNeRF improved boundary sharpness, particularly at wall-ceiling transitions. MVSNeRF captured geometric consistency effectively, with shapes closer to reality and improved detail recovery, though textures on small objects remained blurred. FreeNeRF achieved geometric consistency comparable to MVSNeRF but excelled in restoring details on small objects. Our method outperformed all others, delivering clear and realistic details on small objects on the floor and boundary regions of walls, while also achieving consistent depth and lighting in transitions between walls and ceilings. In the quantitative analysis, we calculated PSNR, SSIM [41], and LPIPS [42] metrics.

It is worth noting that the PixelNeRF and MVSNeRF networks used in the experiments were not fine-tuned for the current scenes. This is because our focus is on reconstruction performance under the three-view condition, while typical fine-tuning procedures require two to three additional images per scene, undermining the comparability of results. Therefore, we used non-fine-tuned versions of these networks for the experiments. The detailed comparison results are presented in Table 3.

Table 3. Quantitative comparison experiments were conducted on the LLFF dataset to evaluate the performance of our network in relation to other methods (↑ indicates higher is better; ↓ indicates lower is better).

Method	PSNR ↑	SSIM ↑	LPIPS ↓
PixelNeRF	7.980	0.289	0.660
DS-NeRF	18.99	0.600	0.420
MVSNeRF	19.30	0.640	0.285
FreeNeRF	19.63	0.690	0.130
Ours	22.56	0.770	0.085

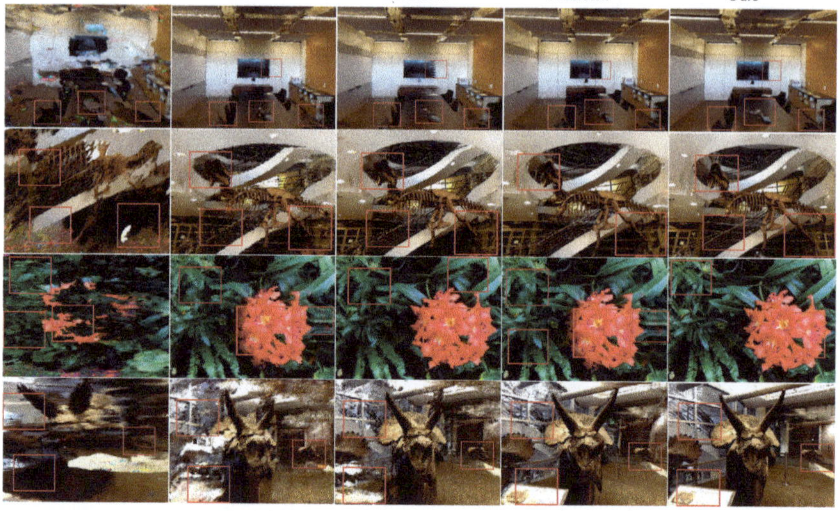

Figure 5. Comparative experiments were conducted on the LLFF dataset to evaluate the performance of our network against four other networks. The results demonstrate that our network achieved superior performance in this benchmark, validating its effectiveness. The red boxes highlight regions where the differences among rendering results of different methods are most noticeable.

4.3. Sparse Input-Based Reconstruction and View Synthesis in Real-World Scenes

In real-world scenarios, accurate camera pose information is typically unavailable, unlike the data provided by synthetic datasets. To achieve precise 3D scene reconstruction and novel view synthesis under these conditions, the pose estimation module of the network must be capable of accurately estimating camera poses even with sparse input. We sampled three real-world scenes using a smartphone and extracted frames at fixed intervals with FFmpeg to create the training and testing images required for our experiments. Our goal is to enable NeRF-based 3D reconstruction and novel view synthesis with only three images in real-world applications.

Experimental results indicate that when other networks employ the traditional SfM algorithm for pose estimation following NeRF's pipeline, the reconstruction quality of all four baseline networks deteriorates significantly. Consequently, a direct comparison with our method becomes impractical. The primary reason lies in the inherent limitations of traditional SfM algorithms, which require substantial overlap between input images and sufficient texture features within overlapping regions. Additionally, camera positions must have a baseline (viewpoint difference) allowing triangulation constraints. Even under these conditions, SfM-estimated camera poses often contain significant noise, leading to poor reconstruction quality and making comparative experiments unnecessary.

Given these limitations, our final experimental setup uses our network's frontend, MASt3R-SfM, to estimate camera poses for all three input images. The same estimated poses are then used for NeRF reconstruction by the four baseline networks. In our assumed real-world scenario (only three images available), there is no standardized test set like the Real Forward-Facing dataset. Ground truth images for quantitative evaluation are unavailable.

To mitigate this, we use adjacent frames near the rendered viewpoints as the best possible reference. We acknowledge that this introduces a slight viewpoint parallax, which can affect the absolute values of the reported metrics. However, since this evaluation condition is applied consistently across all compared methods, it provides a fair and valid basis for assessing their relative performance. We render images from the same viewpoints across all networks for comparison. The ground truth consists of adjacent frames near rendered viewpoints, which—despite slight viewpoint differences—represent the best possible reference setup.

To mitigate viewpoint differences in the ground truth during quantitative experiments, we also render depth maps for each network. These depth maps provide direct visualization of geometric consistency and detail recovery. The ground truth depth maps are estimated using the current state-of-the-art monocular depth estimation network, DepthAnythingV2 [43,44]. It is important to note that these depth maps serve strictly as a qualitative visual reference for geometric plausibility and are not used for any quantitative evaluation metrics. Our quantitative analysis is performed exclusively on the rendered RGB images. In these maps, closer points are shown in white and farther points in black—the inverse of NeRF-rendered depth maps, where closer points appear in black and farther points in white. The qualitative and quantitative experimental results are shown in Figure 6 and Table 4, respectively.

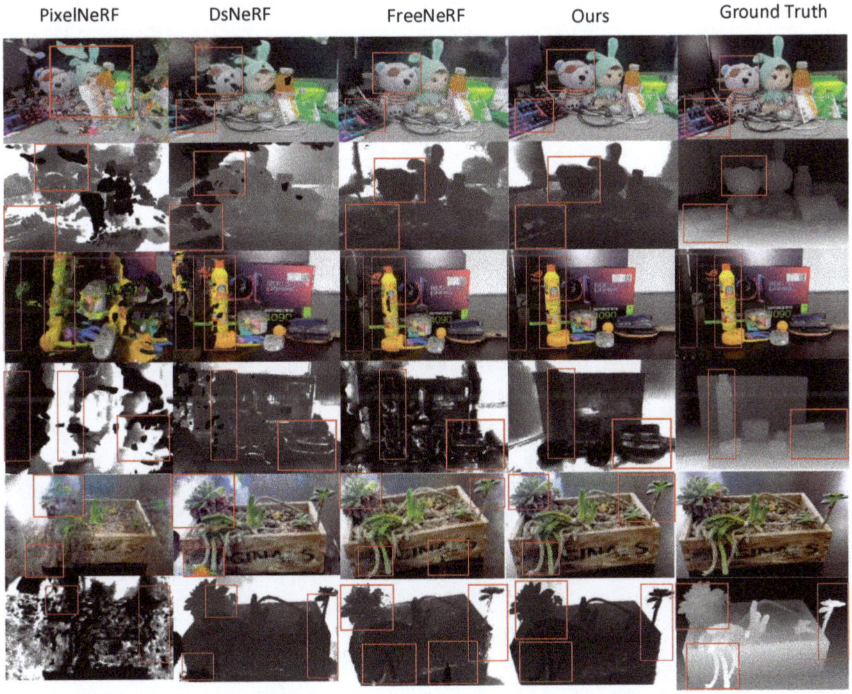

Figure 6. Three-view 3D reconstruction and novel view synthesis from smartphone-captured images. This experiment uses images captured by a smartphone and compares the performance of different sparse-view methods under the same camera poses estimated by our frontend. The figure alternately displays the rendered RGB images and their corresponding depth maps. It is evident that other baseline networks (e.g., PixelNeRF, DSNeRF, and FreeNeRF) exhibit significant defects in both geometric reconstruction (depth maps) and appearance rendering (RGB images), such as structural collapse, blurry edges, and distorted textures. In contrast, our method (Ours) performs best in both aspects, achieving the most complete geometric reconstruction and rendering the clearest, most realistic images. The red boxes highlight regions where rendering differences among different methods are most evident. The non-English texts visible on certain objects (e.g., packaging labels) are part of the original captured scenes.

Table 4. Quantitative comparison experiments were conducted across three distinct scenarios (↑ indicates higher is better; ↓ indicates lower is better).

Method	PSNR ↑	SSIM ↑	LPIPS ↓
PixelNeRF	6.482	0.236	0.751
DS-NeRF	16.31	0.436	0.643
FreeNeRF	17.35	0.536	0.436
Ours	20.47	0.732	0.096

From the results of quantitative experiments, it is evident that our model achieves superior reconstruction quality compared to other sparse-view NeRF networks. This advantage primarily stems from the incorporation of the sampling annealing strategy and a series of regularization optimization techniques in our network. Our method demonstrates strong performance across three quantitative metrics. From the RGB images output by different comparison networks, it is apparent that PixelNeRF produces subpar reconstruction quality (MVSNeRF performs at a similar level and is therefore not presented separately), exhibiting significant blurriness and artifacts. Many regions lose texture and color information, likely due to the lack of fine-tuning, which substantially degrades the reconstruction quality of these networks. In the red-boxed areas, there are noticeable color shifts and texture blurriness (e.g., the doll's face, the leaves of the plant, and the bottom box). Although DSNeRF shows notable improvements compared to PixelNeRF, it still suffers from artifacts and color shifts. The occlusion boundaries in the red-boxed areas (e.g., the edges of the doll's ears and the plant's leaves) exhibit blending issues, leading to unnatural texture and color rendering. FreeNeRF achieves better texture restoration in the red-boxed regions (e.g., plant details) but still encounters blurriness in geometrically complex regions (e.g., the edges of plant leaves and the doll's ears). In contrast, our method accurately reconstructs texture and color in the red-boxed regions (e.g., the doll's ears, plant leaves, and bottles) without noticeable blurriness or artifacts. The occlusion boundaries and segmentation between the foreground and background are sharp, with high color consistency. Our network also reconstructs fine textures in complex geometric areas (e.g., plant stems and leaves) with minimal artifacts. Additionally, comparing the depth maps output by different networks reveals that PixelNeRF produces depth maps with significant blurriness and noise, particularly in detail-rich or geometrically complex regions (e.g., areas marked by red boxes). Although DSNeRF demonstrates significant improvement over PixelNeRF, it still exhibits incomplete geometry in certain regions and insufficient sharpness at depth boundaries. FreeNeRF generally outperforms DSNeRF in sharpness, but depth estimation errors persist in some regions (e.g., gaps between objects or small objects). Our method produces depth maps with the best clarity, detail, and boundary sharpness, closely approximating the ground truth. The red-boxed regions (e.g., edge details, occluded areas, and small objects) demonstrate significantly higher accuracy than other methods. Overall, our approach outperforms other methods significantly in both RGB rendering and depth map reconstruction quality.

4.4. Ablation Study

4.4.1. Efficacy of the High-Frequency Annealing Strategy

To isolate and rigorously assess the contribution of our high-frequency annealing strategy for positional encoding in sparse-view scenarios, we conduct a targeted ablation study. This evaluation is performed on the NeRF-synthetic dataset, as its scenes—characterized by sharp geometric boundaries and uniform color surfaces—present an ideal testbed for evaluating the network's susceptibility to overfitting on high-frequency information. To eliminate confounding factors from camera pose inaccuracies, we utilize the ground truth camera poses provided with the dataset. This experimental setup ensures that any observed performance differences are solely attributable to the inclusion or exclusion of the proposed annealing strategy.

We compare the following two model variants, each trained using three sparse input views.

Ours (w/Annealing): This is our full model, which incorporates the proposed high-frequency annealing strategy. The strategy progressively unmasks frequency bands by incrementally increasing the maximum level, Lmax, of the positional encoding. This compels the network to first learn the coarse, low-frequency structure of the scene during the initial training stages before gradually enabling the learning of fine, high-frequency details.

Ours (w/o Annealing): This is a baseline variant where the high-frequency annealing strategy is ablated. In this configuration, the network has access to the full spectrum of positional encoding frequencies from the outset of training. This approach, particularly in a sparse-view setting, makes the model prone to overfitting on high-frequency noise rather than capturing the underlying global scene structure.

The experimental results clearly demonstrate the necessity of the high-frequency annealing strategy for mitigating overfitting in sparse-view settings. As illustrated in Figure 7, the baseline model (top row), which lacks this strategy, exhibits catastrophic failure. Specifically, its rendered novel views are plagued by erroneous geometric artifacts, often termed "floaters," that appear in otherwise empty regions of space, resulting in an overall blurry and noisy scene. In sharp contrast, our full model (bottom row) with the annealing strategy generates clean, sharp, and geometrically faithful images that are highly consistent with the ground truth. This stark visual disparity is corroborated by the quantitative metrics in Table 5. Our full model achieves a PSNR of 24.47, more than doubling the 12.31 score of the baseline. Similarly, it shows a decisive advantage across other metrics, with SSIM improving from 0.513 to 0.832 and LPIPS decreasing from 0.482 to 0.162. These results confirm substantial improvements in both structural similarity and perceptual quality.

Figure 7. Qualitative ablation study of the high-frequency annealing strategy. On the NeRF-synthetic dataset, removing this strategy (**top row**) leads to severe overfitting, manifested as significant geometric "floaters" in empty space. Our full model (**bottom row**) successfully suppresses these artifacts, producing clean and sharp images.

Table 5. Quantitative results of the ablation study on high-frequency annealing. (↑ indicates higher is better; ↓ indicates lower is better).

Method	PSNR ↑	SSIM ↑	LPIPS ↓
Ours (w/o Annealing)	12.31	0.513	0.482
Ours (w/Annealing)	24.47	0.832	0.162

This experiment conclusively demonstrates that our high-frequency annealing is a critical component for robust reconstruction from sparse views. By guiding the network to learn in a coarse-to-fine manner, it ensures the model first establishes a robust, low-frequency foundation of the scene before progressively refining high-frequency details. Notably, even under the idealized condition of perfect camera poses, this strategy remains indispensable for achieving high-fidelity renderings.

4.4.2. Impact of Low-Frequency Regularization on Geometric Accuracy

To verify the synergistic contribution of our two-part regularization strategy—which involves sampling contiguous patches from low-frequency regions and an accompanying depth regularization term, L_U, based on the Pearson correlation coefficient—we conduct a targeted ablation study on the DTU dataset. This strategy is designed to stabilize the training process and enhance geometric accuracy. The DTU dataset is selected for its primary advantage of providing high-precision, ground truth 3D geometry, which enables a direct quantitative evaluation of our model's geometric fidelity. To isolate the impact of the regularization strategy, we utilize the provided ground truth camera poses, thereby eliminating potential confounding effects from pose estimation errors.

We compare two variants of our model, both trained on sparse input views:

Ours (Full Model): Our standard model, which employs both the low-frequency patch sampling strategy and the Pearson correlation-based depth loss (L_U) during the first half of the training process ($t \leq T/2$).

Ours (w/o Low-Freq. Regularization): A baseline variant where both components of the low-frequency regularization are ablated. In this setting, the sampling scheme reverts to standard random pixel sampling across the entire image, and the depth regularization term LU is omitted from the loss function.

Our ablation study reveals that the low-frequency regularization strategy is critical for ensuring geometric stability and the final rendering quality. This is particularly evident in the qualitative results presented in Figure 8. The baseline model (top row), which omits this strategy, suffers from severe visual artifacts. Specifically, its rendered surfaces exhibit unnatural splotches of color and shading, an effect typically indicative of an incorrectly learned underlying 3D geometry. Furthermore, the boundaries between the object and the background are poorly defined, further evidencing defects in its geometric boundary reconstruction.

These qualitative degradations are quantitatively substantiated by the metrics in Table 6. Our full model achieves a PSNR of 21.63, outperforming the baseline's 15.23 by a significant margin of 6.4 points. Commensurate advantages are observed in the SSIM and LPIPS metrics, collectively demonstrating a substantial leap in both image fidelity and perceptual quality. We also note that the training process of the baseline is markedly unstable, exhibiting slower convergence and greater volatility in its loss curve.

Table 6. Quantitative results of the ablation study on low-frequency regularization. (↑ indicates higher is better; ↓ indicates lower is better).

Method	PSNR ↑	SSIM ↑	LPIPS ↓
Ours (w/o Low-Freq. Regularization)	15.23	0.693	0.305
Ours (w/Low-Freq. Regularization)	21.63	0.849	0.186

In summary, both the qualitative and quantitative results provide compelling evidence for our central claim: the strategy of sampling contiguous patches in low-frequency regions and the Pearson-based depth regularization loss are complementary and act in synergy. This composite strategy plays a pivotal role throughout the optimization process by stabilizing training and ensuring geometric integrity. It is therefore indispensable for mitigating geometric artifacts induced by sparse views to ultimately achieve high-fidelity view synthesis.

Figure 8. Qualitative ablation study of the low-frequency regularization strategy. On the DTU dataset, removing this strategy (**top row**) results in blurry renders with distorted lighting and structural errors. Our full model (**bottom row**) reconstructs sharp and geometrically correct views, proving the strategy's importance for render quality.

4.4.3. Ablation on the Front-End Pose and Geometry Estimation

To demonstrate the foundational and indispensable role of our integrated MASt3R-SfM front-end, we conduct a final ablation study. We establish a baseline that emulates a conventional NeRF pipeline by first estimating camera poses from the sparse input views using a classical SfM algorithm (i.e., COLMAP). Replacing our MASt3R-SfM front end, however, deprives the system of the initial self-supervised depth maps and the final normalized point cloud it generates. Consequently, our core depth regularization losses, LU and LM, cannot be applied.

To enable this baseline to function, we leverage a pre-trained DepthAnythingV2 model to generate monocular relative depth priors for each input view. These priors are then used to guide our patch sampling strategy and to offer a basic geometric scaffold. This comparative experiment is performed on the challenging forward-facing scenes of the LLFF dataset.

Our final ablation study conclusively demonstrates that the integrated MASt3R-SfM front-end is the cornerstone of our framework. As shown in Figure 9 and Table 7, replacing it with a baseline composed of traditional SfM and a monocular depth estimator leads to a precipitous and catastrophic decline in reconstruction performance. The baseline's renderings are rife with severe geometric distortions and blurry artifacts, rendering the scene structure entirely unrecognizable. This visual collapse is corroborated by the quantitative metrics: the baseline achieves a PSNR of only 7.56, a score indicative of complete reconstruction failure. In stark contrast, our full pipeline achieves a PSNR of 21.36, with a similarly vast performance gap observed for the SSIM and LPIPS metrics.

Figure 9. Qualitative ablation study of the frontend pose and geometry estimation module. On the LLFF dataset, replacing our frontend with traditional SfM (**top row**) results in catastrophic failure, with both RGB images and depth maps showing a total collapse of the scene structure. Our full pipeline (**bottom row**) successfully reconstructs a coherent and high-quality scene.

Table 7. Quantitative results of the ablation study on the frontend module. (↑ indicates higher is better; ↓ indicates lower is better).

Method	PSNR ↑	SSIM ↑	LPIPS ↓
Ours (w/o Annealing)	7.56	0.232	0.693
Ours (w/Annealing)	21.36	0.770	0.264

The root cause of this failure is twofold. First, for complex, real-world scenes such as those in the LLFF dataset, traditional SfM algorithms struggle to estimate accurate and robust camera poses from only 3–5 sparse views. The resulting pose inaccuracies fundamentally corrupt the geometric foundation required for NeRF. Second, while the monocular depth maps from DepthAnythingV2 provide relative depth cues, they fall significantly short of the priors generated by our MASt3R-SfM in terms of accuracy, scale, and multi-view consistency.

Ultimately, this experiment demonstrates that MASt3R-SfM is more than a mere pose estimator; it provides a dual-pronged foundation for the system. It delivers not only high-accuracy, robust camera poses but also a set of multi-view consistent geometric priors that are critical for regularization. Under the challenging conditions of extreme view sparsity in real-world scenes, both of these elements are indispensable prerequisites for achieving high-quality reconstruction.

Taken together, these three ablation studies demonstrate the strong synergistic effects between our proposed components. The front-end ablation (Section 4.4.3) shows that the robust pose estimation and multi-view consistent geometry from MASt3R-SfM form an indispensable foundation. Concurrently, the back-end ablations (Sections 4.4.1 and 4.4.2) prove that even with a strong geometric foundation, our novel regularization strategies are critical for preventing overfitting and achieving high-fidelity results under sparse-view conditions.

5. Discussion

Our research is predicated on a key hypothesis: under real-world sparse-view conditions, front-end camera pose estimation constitutes a more fundamental and fragile bottleneck than back-end NeRF rendering. Our experimental results, particularly the front-end ablation study (Figure 9, Table 7), provide conclusive evidence for this hypothesis. When we replaced our front end with a traditional SfM pipeline, the entire reconstruction system suffered a catastrophic failure even while the back-end NeRF architecture remained unchanged. This finding reveals a critical issue often sidestepped in prior work that assumes known or accurate camera poses.

This result positions our work in dialog with existing literature. On one hand, compared to sparse-view methods dedicated to improving NeRF regularization (e.g., DS-NeRF and FreeNeRF), our work advances this line of research. We argue that in practical sparse-view scenarios, the "accurate pose" premise upon which these methods are built often does not hold. Therefore, SparsePose–NeRF is not merely a regularization scheme but a more holistic solution, spanning from raw images to the final model. By integrating MASt3R-SfM, we first address the "elephant in the room"—pose estimation—thereby allowing the back-end regularization strategies to realize their full potential on a stable and reliable foundation.

On the other hand, in contrast to pose-free methods that attempt to jointly optimize poses and the scene (e.g., BARF, NoPe–NeRF), our framework adopts a distinct and more pragmatic philosophy: decoupling and specialization. While the concept of joint optimization is elegant, its vast search space renders it highly susceptible to local minima under extreme view sparsity. Our results demonstrate that by first resolving the geometry with a specialized and battle-tested front end, the learning burden on the back-end NeRF is substantially reduced. This "divide and conquer" approach exhibits superior robustness and practicality in challenging sparse-view conditions. Regarding the robustness to image noise, our framework has two inherent advantages. First, the learning-based MASt3R-SfM front-end, trained on diverse real-world imagery, is less sensitive to common image noise and blur compared to classical keypoint-based SfM. Second, our back-end regularization, particularly the high-frequency annealing strategy, explicitly prevents the network from overfitting to high-frequency details, which would include sensor noise, thereby promoting a smoother and more plausible scene representation.

In summary, our work highlights the immense potential of a deep integration between classic multi-view geometry and modern neural rendering. This hybrid paradigm, combining the strengths of classical geometry and neural rendering, offers a valuable blueprint for tackling other complex vision tasks, such as robotic navigation and scene understanding.

6. Conclusions and Future Work

6.1. Conclusions

In this paper, we introduced SparsePose–NeRF, a highly robust 3D reconstruction framework designed to address the dual challenges of sparse views and unknown camera poses, which are prevalent in real-world applications of Neural Radiance Fields (NeRF). Conventional NeRF pipelines falter when processing such non-ideal data, as their front-end SfM algorithms often fail due to insufficient information. To overcome this hurdle, our core contribution is the introduction of the advanced MASt3R-SfM algorithm as a powerful front end. It not only robustly estimates high-accuracy camera poses from extremely sparse views but also provides the subsequent NeRF reconstruction with invaluable multi-view consistent geometric priors.

Building on this foundation, we further designed novel back-end regularization strategies. By employing high-frequency annealing for positional encoding, we effectively guide the network to learn scenes in a coarse-to-fine manner, thus preventing overfitting to high-frequency noise under data sparsity. Through an innovative two-stage depth loss function—comprising a local, Pearson correlation-based loss (L_U) and a global, confidence-based loss (L_M)—we significantly enhanced the model's perception of and constraints on the 3D geometric structure. Extensive comparative experiments and ablation studies demonstrate that our method's reconstruction quality comprehensively surpasses that of existing state-of-the-art methods across various datasets, particularly on the most challenging real-world scenes (LLFF). Ultimately, our work represents a solid step towards transitioning NeRF technology from idealized laboratory settings to complex and varied real-world applications.

26. Yang, J.; Pavone, M.; Wang, Y. Freenerf: Improving few-shot neural rendering with free frequency regularization. In Proceedings of the IEEE/CVF Conference on Computer Vision and Pattern Recognition, Vancouver, BC, Canada, 17–24 June 2023; pp. 8254–8263.
27. Wang, Q.; Wang, Z.; Genova, K.; Srinivasan, P.P.; Zhou, H.; Barron, J.T.; Martin-Brualla, R.; Snavely, N.; Funkhouser, T. Ibrnet: Learning multi-view image-based rendering. In Proceedings of the IEEE/CVF Conference on Computer Vision and Pattern Recognition, Nashville, TN, USA, 20–25 June 2021; pp. 4690–4699.
28. Deng, K.; Liu, A.; Zhu, J.-Y.; Ramanan, D. Depth-supervised nerf: Fewer views and faster training for free. In Proceedings of the IEEE/CVF Conference on Computer Vision and Pattern Recognition, New Orleans, LA, USA, 18–24 June 2022; pp. 12882–12891.
29. Roessle, B.; Barron, J.T.; Mildenhall, B.; Srinivasan, P.P.; Nießner, M. Dense depth priors for neural radiance fields from sparse input views. In Proceedings of the IEEE/CVF Conference on Computer Vision and Pattern Recognition, New Orleans, LA, USA, 18–24 June 2022; pp. 12892–12901.
30. Jain, A.; Tancik, M.; Abbeel, P. Putting nerf on a diet: Semantically consistent few-shot view synthesis. In Proceedings of the IEEE/CVF International Conference on Computer Vision, Montreal, BC, Canada, 10–17 October 2021; pp. 5885–5894.
31. Wang, C.; Chai, M.; He, M.; Chen, D.; Liao, J. Clip-nerf: Text-and-image driven manipulation of neural radiance fields. In Proceedings of the IEEE/CVF Conference on Computer Vision and Pattern Recognition, New Orleans, LA, USA, 18–24 June 2022; pp. 3835–3844.
32. Yu, A.; Ye, V.; Tancik, M.; Kanazawa, A. pixelnerf: Neural radiance fields from one or few images. In Proceedings of the IEEE/CVF Conference on Computer Vision and Pattern Recognition, Nashville, TN, USA, 20–25 June 2021; pp. 4578–4587.
33. Chen, A.; Xu, Z.; Zhao, F.; Zhang, X.; Xiang, F.; Yu, J.; Su, H. Mvsnerf: Fast generalizable radiance field reconstruction from multi-view stereo. In Proceedings of the IEEE/CVF International Conference on Computer Vision, Montreal, BC, Canada, 10–17 October 2021; pp. 14124–14133.
34. Leroy, V.; Cabon, Y.; Revaud, J. Grounding image matching in 3d with mast3r. In Proceedings of the European Conference on Computer Vision, Milan, Italy, 29 September–4 October 2024; pp. 71–91.
35. Wang, S.; Leroy, V.; Cabon, Y.; Chidlovskii, B.; Revaud, J. Dust3r: Geometric 3d vision made easy. In Proceedings of the IEEE/CVF Conference on Computer Vision and Pattern Recognition, Seattle, WA, USA, 16–22 June 2024; pp. 20697–20709.
36. Knapitsch, A.; Park, J.; Zhou, Q.-Y.; Koltun, V. Tanks and temples: Benchmarking large-scale scene reconstruction. *ACM Trans. Graph. (ToG)* **2017**, *36*, 1–13. [CrossRef]
37. Reizenstein, J.; Shapovalov, R.; Henzler, P.; Sbordone, L.; Labatut, P.; Novotny, D. Common objects in 3d: Large-scale learning and evaluation of real-life 3d category reconstruction. In Proceedings of the IEEE/CVF International Conference on Computer Vision, Montreal, BC, Canada, 10–17 October 2021; pp. 10901–10911.
38. Zhou, T.; Tucker, R.; Flynn, J.; Fyffe, G.; Snavely, N. Stereo magnification: Learning view synthesis using multiplane images. *arXiv* **2018**, arXiv:1805.09817. [CrossRef]
39. Sturm, J.; Engelhard, N.; Endres, F.; Burgard, W.; Cremers, D. A benchmark for the evaluation of RGB-D SLAM systems. In Proceedings of the 2012 IEEE/RSJ International Conference on Intelligent Robots and Systems, Vilamoura-Algarve, Portugal, 7–12 October 2012; pp. 573–580.
40. Mildenhall, B.; Srinivasan, P.P.; Ortiz-Cayon, R.; Kalantari, N.K.; Ramamoorthi, R.; Ng, R.; Kar, A. Local light field fusion: Practical view synthesis with prescriptive sampling guidelines. *ACM Trans. Graph. (ToG)* **2019**, *38*, 1–14. [CrossRef]
41. Wang, Z.; Bovik, A.C.; Sheikh, H.R.; Simoncelli, E.P. Image quality assessment: From error visibility to structural similarity. *IEEE Trans. Image Process.* **2004**, *13*, 600–612. [CrossRef] [PubMed]
42. Zhang, R.; Isola, P.; Efros, A.A.; Shechtman, E.; Wang, O. The unreasonable effectiveness of deep features as a perceptual metric. In Proceedings of the IEEE Conference on Computer Vision and Pattern Recognition, Salt Lake City, UT, USA, 18–23 June 2018; pp. 586–595.
43. Yang, L.; Kang, B.; Huang, Z.; Xu, X.; Feng, J.; Zhao, H. Depth anything: Unleashing the power of large-scale unlabeled data. In Proceedings of the IEEE/CVF Conference on Computer Vision and Pattern Recognition, Seattle, WA, USA, 16–22 June 2024; pp. 10371–10381.
44. Yang, L.; Kang, B.; Huang, Z.; Zhao, Z.; Xu, X.; Feng, J.; Zhao, H. Depth Anything V2. *arXiv* **2024**, arXiv:2406.09414. [CrossRef]

photonics MDPI

Article

Mathematical Analysis and Freeform Surface Modeling for LED Illumination Systems Incorporating Diffuse Reflection and Total Internal Reflection

Xin Xu [1,*], Jianghua Rao [1], Xiaowen Liang [1], Zhenmin Zhu [2] and Yuanyuan Peng [2]

[1] School of Mechanical and Electronic Engineering, Gandong University, Fuzhou 344000, China
[2] School of Electrical and Automation Engineering, East China Jiaotong University, Nanchang 330013, China; 3066@ecjtu.edu.cn (Y.P.)
* Correspondence: xinxu@gdc.edu.cn

Abstract

Indirect lighting systems employing light-emitting diodes (LEDs) and diffuse reflective surfaces are prevalent in applications demanding stringent illumination uniformity. However, conventional diffuse reflection approaches exhibit inherent limitations (inevitable light loss from multiple diffuse reflections and trade-off between uniformity and efficiency). To overcome these constraints, we introduce a novel composite freeform surface illumination system that synergistically integrates total internal reflection (TIR) with diffuse reflection. This design leverages the inherent Lambertian radiation characteristics of LEDs and the properties of ideal diffuse reflectors. A rigorous mathematical model is derived based on the luminous intensity distribution of the LED chip, the prescribed illumination requirements on the target plane, the principle of energy conservation, and Snell's law. The resulting system of nonlinear equations is solved to generate a series of two-dimensional profile curves, which are subsequently synthesized into an off-axis freeform surface. Simulated results demonstrate that the proposed system achieves higher optical efficiency and superior illumination uniformity compared to traditional diffuse reflector configurations. This universal and feasible methodology broadens the application potential of high-performance diffuse indirect lighting.

Keywords: LED; optical design; diffuse reflection; freeform surface; uniform illumination

1. Introduction

As the fourth-generation lighting source, LEDs offer several advantages, including compact size, rapid response, long lifespan, high brightness, and low thermal output [1–5]. Therefore, they have not only transformed people's perception of general lighting design and application but also provided key solutions to address the needs for energy conservation and environmental protection in the present and future. In optical design, a freeform surface is strictly defined as a non-symmetric surface (without rotational or translational symmetry) whose sag (surface profile) cannot be completely described by simple mathematical functions such as conic sections, spherical equations, or rotational symmetric polynomials. This distinguishes it from rotationally symmetric surfaces (e.g., spherical, paraboloidal) and simplified symmetric surfaces, whose profiles rely on symmetric geometric constraints. Due to the approximately Lambertian cosine distribution of the spatial luminous intensity of LEDs, direct illumination without secondary light distribution can lead to light pollution and non-uniform illumination [6]. Therefore, in the practical application of LED lighting, secondary light distribution design is essential for LEDs to ensure that their light energy distribution meets practical lighting requirements [7–9]. In recent years, researchers have focused on designing double freeform surface lenses to achieve uniform illumination on the target plane, but this approach is also accompanied by a series of issues such as overexposure and glare effects caused by direct and simple reflective structures. So the indirect lighting technology based on diffuse reflective freeform surfaces has emerged [10,11]. Diffuse reflection technology can also be used to enhance the illumination uniformity of lighting systems. B. Liu proposed an indirect illumination method, which is composed of planar LED array and diffuse reflection freeform surface, the simulation result shows that the irradiance uniformity of the detection area is above 90% [12]. CH Tsuei designed an LED lighting system for indoor lighting using diffuse reflectors. Their research not only improved the uniformity of the target illuminated surface but also suppressed glare effects to some extent [13]. So illumination uniformity can also be enhanced through diffuse reflection. However, the diffuse reflectors are less efficient than other optical designs in terms of light efficiency. Therefore, improving the efficiency of diffuse reflectors remains an important research focus. Zhu incorporated diffuse reflection into the design of lighting systems, establishing a system of nonlinear equations based on LED circular arrays and high diffuse reflection surfaces [14]. Solving this system of nonlinear equations via relevant algorithms yields both the coordinate data and contour of the diffuse reflection freeform surface. The required freeform surface shape is then obtained through rotational symmetry about the axis. Simulation experiments have demonstrated that, compared with traditional direct lighting, this design performs well in terms of irradiation uniformity and lighting efficiency. However, the theory underlying diffuse reflection still requires refinement. To address this, Zhu et al. implemented energy-saving measures, which significantly mitigated the low efficiency of the design. They established a mathematical simulation algorithm for the radiation of the illuminated plane using the bidirectional reflectance distribution function (BRDF) of the inner surface of the diffuse reflection freeform surface. Based on conservation laws, they calculated the expected values on the target plane and established a set of nonlinear algebraic equations. Compared with traditional design methods (direct illumination and hemispherical inner surface diffuse illumination), this method exhibits better uniformity and illumination efficiency [15]. Sun Xiang proposed an off-axis design method for diffuse reflective surface. Compared with conventional approaches, this method enables the generation of distinct asymmetric surfaces tailored to different LED arrays under far-field conditions, thereby achieving both enhanced efficiency and significantly improved uniformity [16].

This paper presents an LED lighting system (a compact optical unit integrating a single LED source and composite freeform surfaces, which is designed to achieve uniform illumination on the target plane) that integrates specular total internal reflection (TIR) and diffuse reflection freeform surfaces. A mathematical model for the irradiation distribution of the freeform surface is established, which is derived from the characteristics of a single LED Lambertian source and the properties of ideal diffuse reflection. Moreover, the desired illumination distribution over the target plane is specified. By incorporating the principles of energy conservation and spatial Snell's law, and based on the collinearity condition of light propagation and the perpendicularity between adjacent point normals and line vectors, a coordinate recursive equation system is established. The generatrix profile of the rotationally symmetric TIR surface is derived via iterative fitting. And then a mathematical simulation algorithm is established to represent the radiation of the irradiated plane by using the bidirectional reflectance distribution function of the diffuse freeform surface. By solving the derived system of nonlinear algebraic equations, a series of 2-D contour curves are obtained and combine these 2-D curves are further synthesized into a diffuse reflection freeform surface. The results show that, in comparison with traditional diffuse reflectors, the proposed illumination system achieves significantly higher illumination efficiency and superior uniformity.

2. Materials and Methods

The design methodology flowchart is presented in Figure 1. In summary, we first conceptualized an LED illumination system incorporating a composite freeform surface that synergizes specular total internal reflection (TIR) with diffuse reflection. Subsequently, a freeform lens was computationally designed based on this concept. The optical performance of the designed lens was rigorously evaluated through Monte Carlo ray-tracing simulations. This design-evaluation process was iterated until the simulation results met the predefined illumination specifications, yielding an optimized freeform lens. The detailed steps of the design methodology are elaborated below.

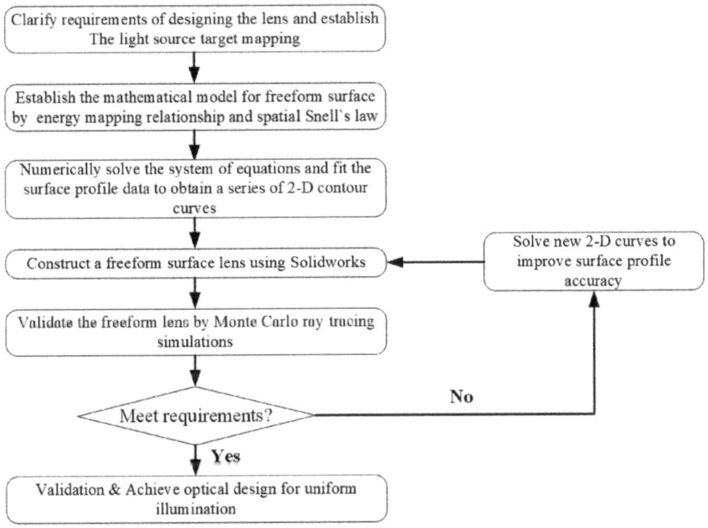

Figure 1. Flowchart of this paper design method for LED uniform illumination.

Figure 2 depicts the cross-sectional profile of the composite freeform surface illumination system developed in this study. Incident light from the LED source is partitioned into two distinct optical paths: (1) rays propagating towards the rotationally symmetric TIR surface undergo total internal reflection and are subsequently directed onto the diffuse reflective freeform surface; (2) rays incident directly upon the diffuse reflective freeform surface propagate towards the target plane. The combined illumination on the target plane arises from the superposition of the light reflected by the diffuse surface (originating from Light Path 1: LED $\rightarrow$ TIR freeform surface $\rightarrow$ diffuse reflective freeform surface $\rightarrow$ target illumination plane) and the light transmitted/reflected by the diffuse surface (originating from Light Path 2: LED $\rightarrow$ diffuse reflective freeform surface $\rightarrow$ target illumination plane).

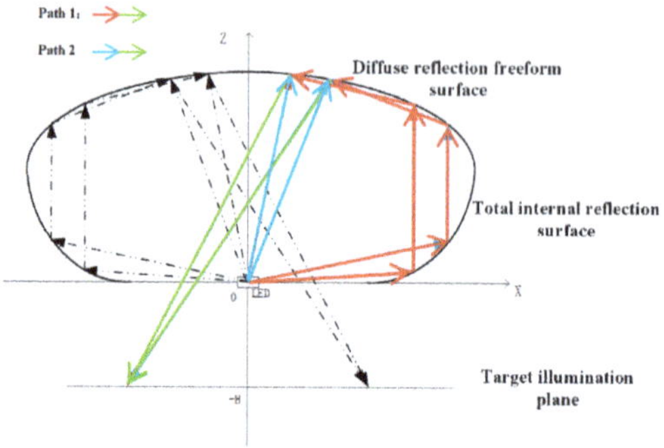

Figure 2. The cross-sectional profile of the combined freeform surface illumination system.

Due to the requirement that the size of the lighting system designed is larger than the diameter of the LED chip, a single LED chip can be approximated as a point light source, and the radiation intensity of a single LED point light source can be obtained:

$$I_e(\theta) = I_0 \cos^m \theta \quad (1)$$

where θ is the angle between the light ray and the axial ray of the LED, $I_e(\theta)$ represents the radiant intensity of a single LED chip at any angle, and I_0 stands for the axial radiant intensity of the LED. Since the emission characteristics of the LED chip can be approximated to a Lambertian source, m is approximately 1. Therefore, the radiant intensity of the entire LED chip can be expressed as:

$$E_e(r, \theta) = \frac{I_0 \cos \theta}{d^2} \quad (2)$$

When transitioning from the spherical coordinate system (Equation (2)) to the Cartesian coordinate system (with the LED chip positioned at the origin), the radial distance d (between the LED and freeform surface points in spherical coordinates) is redefined as S (Euclidean distance in Cartesian coordinates, $S = \sqrt{(X - X_i)^2 + (Z - Z_i)^2}$, where (X_i, Z_i) are the LED coordinates). Substituting $d = S$ and $cos\theta = \frac{Z}{S}$ (since θ is the angle between the light ray and the LED's axial ray, $Z = Scos\theta$) into Equation (2) yields Equation (3):

$$E_e(S, Z) = \frac{Z \cdot I_0}{S^3} \quad (3)$$

$E_e(S, Z)$ denotes the irradiance of each discrete point on the freeform surface of the LED chip, S represents the distance between a point on the diffuse reflection surface and the LED, and the plane of the 2-D curve obtained by iteration is perpendicular to the XOY plane. So Equation (3) can be simplified as:

$$E_e(x, 0, z) = \frac{z \cdot I_0}{\left[(x - x_i)^2 + (z - z_i)^2\right]^{3/2}} \quad (4)$$

(x_i, z_i) denotes the position coordinates of the LED, and Equation (4) represents the irradiance distribution model for a single LED.

2.1. Total Internal Reflection (TIR) Freeform Surface Design

Firstly, a coordinate system is established as shown in Figure 3, with the central axis of the TIR lens designated as the z-axis and the position of the light source set as the coordinate origin [17]. Assuming the LED light source is an ideal point source, light rays emanate directly from the origin and are incident on point $Q_i(x_i, z_i)$ of rotationally symmetric TIR surface. After undergoing total internal reflection, the rays emerge as light rays parallel to the z-axis.

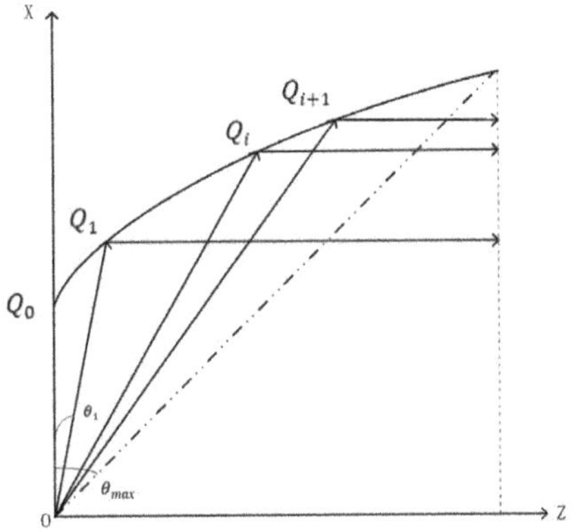

Figure 3. Calculation of total internal reflection freeform surface.

Define the angle between the light emitted by the LED source and the x-axis as θ_i (must be consistent with the TIR surface's collimation constraint), and the maximum incident angle as θ_{max}. Divide θ_{max} evenly into N parts, such that $\theta_i = 90° - i{\cdot}\theta_{max}/N(i = 1, 2, \ldots, N)$. For TIR, the reflection direction of light satisfies the law of reflection (angle of incidence = angle of reflection) and the phase condition derived from Maxwell's equations. The unit vector of incident light inside the lens is $\vec{q_i} = (sin\theta_i, cos\theta_i)(\theta_i$: angle between incident ray and x-axis), and the reflected light (parallel to z-axis) has a unit vector of $\vec{r_i} = (0, 1)$. Based on the vector relationship between incident ray, reflected ray, and surface normal ($\vec{N_i}$) for TIR:

$$n{\cdot}\vec{q_i} + \vec{r_i} = k{\cdot}\vec{N_i} \quad (5)$$

where k is a positive scaling factor (ensuring normal vector direction), n is the lens refractive index, $\vec{q_i}$ is the incident light unit vector, $\vec{r_i}$ is the reflected light unit vector, and $\vec{N_i}$ is the TIR surface normal vector at point Q_i. Substituting $\vec{q_i} = (sin\theta_i, cos\theta_i)$ and $\vec{r_i} = (0, 1)$ into Equation (5), and normalizing the normal vector, we re-derive the correct expression for $\vec{N_i}$ (consistent with the physical meaning of TIR):

$$\vec{N_i} = [-(sin\theta_i + K_i), n - cos\theta_i] \quad (6)$$

K_i is a scaling coefficient for the x-component of the normal vector, defined as $K_i = \frac{n{\cdot}sin\theta_i}{k}$, where $k = \sqrt{n^2 + 2ncos\theta_i + 1}$ (scaling factor from Equation (5)). It quantifies the contribution of incident light's horizontal component to the TIR surface normal direction. The incident light ray is collinear with the unit vector $\vec{q_i}$ from the origin to point $Q_i(x_i, z_i)$. Thus, it can be concluded that:

$$z_i = x_i \cdot \cot\theta_i \quad (7)$$

The line connecting adjacent points Q_i and Q_{i+1} is perpendicular to the normal at point Q_i, that is:

$$(x_{i+1} - x_i) \cdot N_{xi} + (z_{i+1} - z_i) \cdot N_{zi} = 0 \quad (8)$$

Based on the above recursive relationship, starting from the initial point $Q_0(x_0, z_0)$, the coordinates of each point $(Q_1, Q_2, \ldots, Q_n)$ on the generatrix are calculated sequentially.

Through iterative convergence, the complete contour of the generatrix of the total internal reflection freeform surface can be obtained. The 2-D contour plot of the designed total internal reflection freeform surface is shown in Figure 4 (the TIR freeform surface has a radial range of 0–12 mm and an axial height of 5–8 mm, matching the composite lens's compact design).

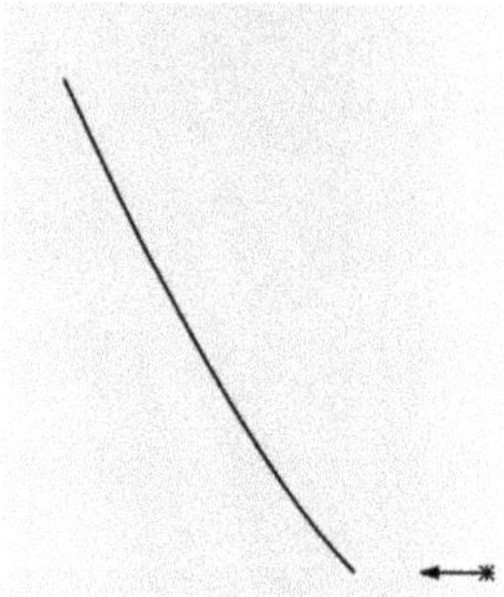

Figure 4. The 2-D contour of the total internal reflection freeform surface.

2.2. *Diffuse Reflection Freeform Surface Design*

The diffuse reflection structure is an off-axis freeform surface (symmetric about the z-axis in Figure 2) with a 3D morphology synthesized from 2D contour curves. Its spatial range spans from the upper edge of the TIR freeform surface (point Q_n in Figure 3) to the outermost boundary of the lens, with a maximum radial width of 35 mm and a height variation of 8–12 mm (optimized to avoid light blockage between the TIR and diffuse surfaces). The surface is discretized into 2000+ discrete points with a spacing of 0.05 mm, ensuring smooth curvature transitions (curvature radius variation < 0.1 mm^{-1}) to prevent local light concentration.

As shown in Figure 5, E_1 denotes the irradiance of light reflected by the mirror surface, E_1' represents the irradiance of light incident on the rotationally symmetric diffuse reflection surface, and E is the total irradiance on the combined freeform surface, we can get:

$$\begin{cases} E_1 = \frac{I_0 \cdot cos\theta}{[\sqrt{x_i^2+z_i^2}+(x-x_i)]^2} \\ E_1' = \frac{I_0 \cdot cos\theta}{x^2+z^2} \\ E = E_1 + E_1' \end{cases} \tag{9}$$

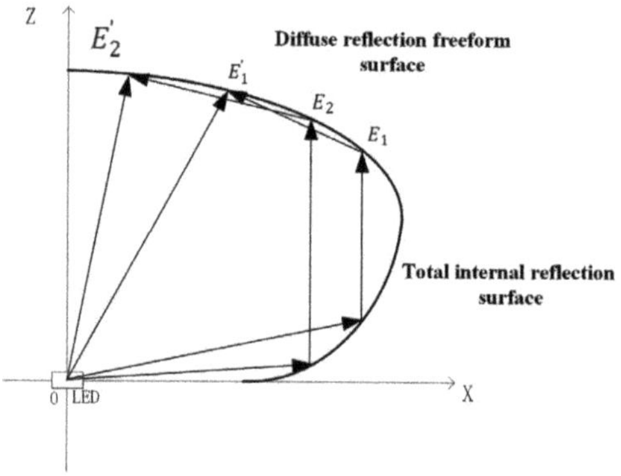

Figure 5. The irradiance of the diffuse reflection freeform surface.

Since the rotationally symmetric diffuse reflection surface is regarded as a series of continuously distributed second-order Lambertian light sources with rotational symmetry, this paper takes the irradiance distribution on the target plane in the X-Z 2-D coordinate system as an example [18]. Figure 6 shows the point P on the rotationally symmetric diffuse reflection surface and point T on the target plane can be represented by coordinate points $P(x_i, 0, z_i)$ and $T(x_t, 0, -H)$ on the 2-D plane. According to the knowledge of space vector, we can derive the following expression:

$$\begin{cases} \overrightarrow{OUT} = (x_t - x, -H - z) \\ \left(\begin{array}{c} \overrightarrow{IN} = (x,z) \\ \overrightarrow{N} = (dz, -dx) \end{array} \right) \\ cos\theta = \dfrac{\overrightarrow{OUT} \cdot \overrightarrow{N}}{\left|\overrightarrow{OUT}\right| \cdot \left|\overrightarrow{N}\right|} \end{cases} \tag{10}$$

where $\overrightarrow{OUT}$ and $\overrightarrow{IN}$ respectively represent the light vector emitted and absorbed from the rotationally symmetric diffuse reflection surface, $\overrightarrow{N}$ is the normal vector on the freeform surface, θ is the angle between $\overrightarrow{OUT}$ and $\overrightarrow{IN}$.

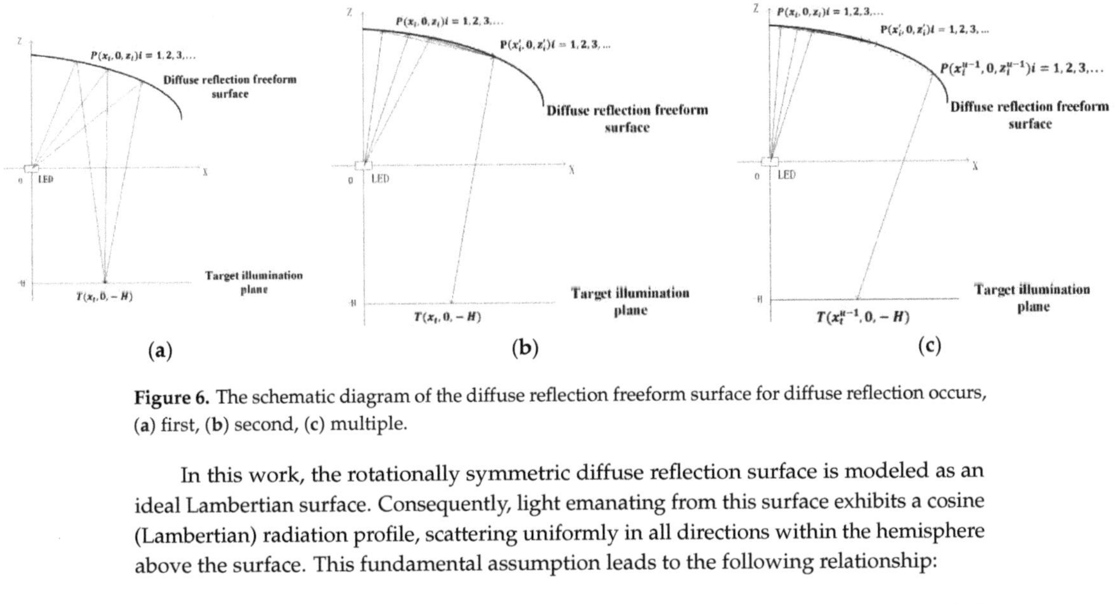

Figure 6. The schematic diagram of the diffuse reflection freeform surface for diffuse reflection occurs, (**a**) first, (**b**) second, (**c**) multiple.

In this work, the rotationally symmetric diffuse reflection surface is modeled as an ideal Lambertian surface. Consequently, light emanating from this surface exhibits a cosine (Lambertian) radiation profile, scattering uniformly in all directions within the hemisphere above the surface. This fundamental assumption leads to the following relationship:

$$I_\theta = I_0 cos\theta \tag{11}$$

where I_0 is the maximum radiance in the direction normal to the rotationally symmetric diffuse reflection surface. Therefore, the irradiance distribution on the target illumination plane can be expressed as:

$$E' = \frac{E \cdot BRDF \cdot cos\theta \cdot dA}{d'^2} \tag{12}$$

where E is the irradiance received by the freeform surface (W/m^2), as given by Equation (4), d' is the distance between the rotationally symmetric diffuse reflection surface and the target illumination plane; and dA denotes the area element on the freeform surface. E' as the general target-plane irradiance formula. The structure is defined as an ideal Lambertian reflector, meaning its bidirectional reflectance distribution function (BRDF) follows the cosine law. This ensures that light reflected from the surface is uniformly scattered within the upper hemisphere (relative to the local surface normal), with no directional dependence. The BRDF describes the reflectance properties of the freeform surface and is given by:

$$BRDF = \frac{\rho}{\pi} \tag{13}$$

where ρ is the rotationally symmetric diffuse reflection surface reflectance, considering the practical scenario, the value of ρ is set to 0.85 in the calculations presented in this article. As shown in Figure 6a, substituting Equations (10)–(13) reduces the irradiance simulation algorithm for point $T(x_t, 0, -H)$ on the target plane, so we can get:

$$E_t(x_t, 0, -H) = \sum_{i=1}^{U} \frac{E \cdot \rho \cdot \left[(x_t - x_i) \cdot dz_i + (H + Z_i) \cdot dx_i\right]}{\pi \cdot \left[(x_t - x_i)^2 + (H + z_i)^2\right]^{3/2} \left[(dz_i)^2 + \left(dx_i\right)^2\right]^{1/2}} \tag{14}$$

Equation (14) quantifies the irradiance on the target plane (at point $T(x_t, 0, -H)$) generated by light that reaches the diffuse reflective freeform surface via initial total internal reflection (TIR) and then undergoes diffuse reflection, where E_t as the specific target-plane irradiance at point T. Correspondingly, as illustrated in Figure 6b, the irradiance distribution on the target illumination plane, generated by the secondary diffuse reflection event from this diffuse reflection surface acting as a Lambertian source, is derived as follows:

$$E'_t(x'_t, 0, -H) = \sum_{i=1}^{U} \frac{E_t(x_t, 0, -H) \cdot \rho \cdot \left[(x'_t - x'_i) \cdot dz_i + (H + z'_i) \cdot dx_i\right]}{\pi \cdot \left[(x'_t - x'_i)^2 + (H + z'_i)^2\right]^{3/2} \left[(dz_i)^2 + \left(dx_i\right)^2\right]^{1/2}} \tag{15}$$

As shown in Figure 6c, point P on the rotationally symmetric diffuse reflection surface and point T on the target plane can be further represented as points $P\left(x_i^{u-1}, 0, z_i^{u-1}\right)$ and $T\left(x_t^{u-1}, 0, -H\right)$ in the X-Z 2-D plane, respectively. This paper assumes that there are N radiation points on the rotationally symmetric diffuse reflection surface, and each radiation point radiates to two non-adjacent points. Therefore, based on the above-described rotationally symmetric diffuse reflection surface, it can be further concluded that when the n-th diffuse reflection occurs, the illuminance distribution of the target illumination plane generated by the diffuse reflection freeform surface is:

$$E_{tu}\left(x_t^{u-1}, 0, -H\right) = \sum_{i=1}^{U} \frac{E_{t(u-1)}\left(x_t^{u-2}, 0, -H\right) \cdot \rho \cdot \left[\left(x_t^{u-1} - x_i^{u-1}\right) \cdot dz_i + \left(H + z_i^{u-1}\right) \cdot dx_i\right]}{\pi \cdot \left[\left(x_t^{u-1} - x_i^{u-1}\right)^2 + \left(H + z_i^{u-1}\right)^2\right]^{3/2} \left[(dz_i)^2 + \left(dx_i\right)^2\right]^{1/2}} \tag{16}$$

The superscripts of coordinates (e.g., $x_i^{u-1}, z_i^{u-1}, x_t^{u-1}$): They represent the spatial coordinates of discrete points on the diffuse surface (x_i, z_i) and target plane corresponding to the u-th diffuse reflection. The $u - 1$ index indicates that these coordinates are updated iteratively based on the previous (($u - 1$)-th) reflection, ensuring consistency with the cumulative effect of multiple reflections. the subscript of irradiance (e.g., E_{tu}): denotes the target-plane irradiance generated by the $u -$ th diffuse reflection on the diffuse surface.

According to the energy conservation constraint, the total radiant flux incident on the diffuse surface (from direct LED rays and TIR-reflected rays) must equal the flux scattered to the target plane (accounting for reflectance losses). Mathematically, this is expressed:

$$\int_{A_{diffuse}} E \cdot dA = \frac{1}{\rho} \int_{A_{target}} E_t \cdot dA \quad (17)$$

where $A_{diffuse}$ is the area of the diffuse surface, E is the irradiance on the diffuse surface (Equation (9)), A_{target} is the target plane area, and E_t is the target irradiance (Equation (14)). This constraint ensures no net energy loss in the reflection process (excluding material absorption). The tangent and normal vectors at the junction between the diffuse surface and the TIR surface (point Q_n) must be continuous to avoid light refraction losses at the interface. This is enforced by setting the 1st and 2nd derivatives of the diffuse surface's 2-D contour (at the junction point) equal to those of the TIR surface's generatrix (Equation (8)).

For each discrete point $P_i(x_i, 0, z_i)$ on the initial contour, the irradiance $E(P_i)$ (Equation (9)) and the corresponding target irradiance $E_t(T_j)$ (Equation (14)) are calculated. The error between the simulated E_t and the desired uniform irradiance ($E_{target,desired} = 100$ lux) is defined as:

$$\delta = \frac{1}{N} \sum_{j=1}^{N} \left| \frac{E_t(T_j) - E_{target,desired}}{E_{target,desired}} \right| \quad (18)$$

where $N = 128 \times 128$ (target plane grid resolution). The contour points are adjusted iteratively using the 4th-order Runge–Kutta algorithm to minimize δ. Each iteration updates the z-coordinate of P_i based on the error gradient ($\frac{\partial \delta}{\partial z_i}$), with a step size of 0.01 mm to ensure convergence. The iteration stops when $\delta < 3\%$ (typically after 50–80 iterations), yielding the optimized 2-D contour (Figure 7a). The optimized 2-D contour is rotated 360° around the z-axis (with 0.5° angular increments) to form the 3-D diffuse reflection surface (Figure 7b).

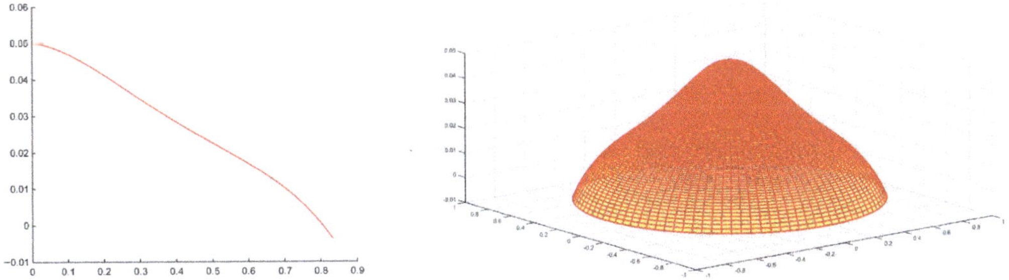

(a) (b)

Figure 7. (**a**) The 2-D contour of the diffuse reflection freeform surface (unit: mm), (**b**) the 3-D diffuse reflection freeform surface.

The computationally designed 2-D contour profiles of the total internal reflection (TIR) freeform surface (derived as shown in Figure 4) and the diffuse reflective freeform surface (derived as shown in Figure 8a) were imported into SolidWorks® (2022 SP3.0) for 3-D reconstruction. The 2D contour of the composite freeform surface (Figure 8a) unifies the TIR and diffuse reflection surface profiles in the z-x plane, where the horizontal axis aligns with the x-axis of Figure 4 (TIR design) and the vertical axis aligns with the z-axis of Figure 3, ensuring consistency with the overall coordinate system. The resultant composite freeform lens geometry is presented in Figure 8 (The composite freeform lens (TIR + diffuse) has a maximum radial width of 25 mm and a volume < 10 cm^3; the 2-D contour in Figure 8a uses a coordinate scale where 1 unit = 1 mm): Figure 8a displays the final 2-D contour profile of the integrated lens design, the 2D contour line of the composite freeform surface (in the z-x plane, with the horizontal axis representing the lateral direction (x-axis) and the vertical axis representing the axial direction (z-axis) of the lens). This contour integrates the TIR freeform surface generatrix (Figure 4) and diffuse reflection freeform surface contour (Figure 7a), serving as the blueprint for 3D reconstruction via rotational symmetry around the z-axis. And Figure 8b shows the corresponding 3D solid model of the fabricated lens component.

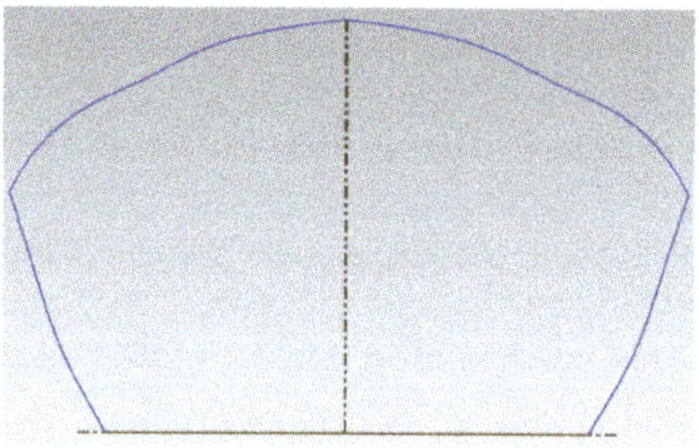

(a)

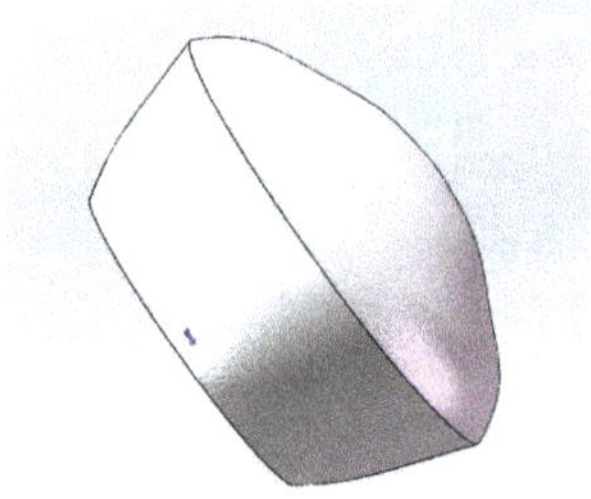

(b)

Figure 8. (**a**) The 2-D contour line of the composite freeform surface, (**b**) the entity model of the lens.

3. Results

Leveraging the design methodology outlined above, a composite freeform optical element integrating total internal reflection (TIR) and diffuse reflection functionalities was developed. Given the scale of the optical system relative to the LED source dimensions, the LED was modeled as a point source exhibiting a Lambertian radiant intensity distribution. The optical power of the LED source was set to 1 W, emitting at a wavelength of 630 nm for initial validation: minimal PMMA absorption ($\geq$92% transmittance), standard for TracePro simulations, ensuring reliable theory-experiment correlation. Though 555 nm (CIE 1931 photopic peak) suits visual lighting, the design's uniformity (>82%) and efficiency (>42%)—dependent on TIR-diffuse synergy—persist across 450–700 nm, aligning better with vision at 555 nm. And The composite freeform optical device is a solid structure fabricated from PMMA (refractive index $n \approx 1.49$). This solid design is essential for realizing the TIR effect and integrating the diffuse reflection layer, ensuring both optical performance and structural stability. The optical performance of the designed system was simulated using TracePro (version 7.8.3) ray-tracing software. The target plane was discretized with a grid resolution of 128 $\times$ 128 pixels, spanning radii from 50 mm to 500 mm. The separation distance between the optical system and the target plane was varied between 150 mm and 1600 mm. To critically evaluate the efficacy and limitations of the proposed design, its simulated lighting efficiency and illumination uniformity were benchmarked against three representative conventional approaches: (i) hybrid diffuse-TIR freeform surface, (ii) diffuse-only freeform surface, and (iii) integrating sphere.

Figure 9 presents the experimental results of the lighting system designed in this paper. It can be observed from the results that when the distance between the light source and the target illumination plane is 400 mm and the radius of the target illumination plane is 100 mm, the irradiance uniformity reaches 83.35%. Table 1 systematically compares the simulated performance metrics of three distinct illumination systems—(i) hybrid diffuse-TIR freeform surface, (ii) diffuse-only freeform surface, and (iii) integrating sphere (standard parameters for small-scale LED indirect illumination testing)—across parametric variations in the LED-to-target-plane distance (50–500 mm) with a constant radius of the target illumination plane (100 mm). A 100 mm-diameter integrating sphere was modeled, with its inner surface coated in a high-reflectivity diffuse material (consistent with commercial models like the Thorlabs IS100 series—Thorlabs, Inc., with its headquarters in Newton, NJ, USA) for LED optical performance characterization and a 20 mm-diameter circular exit aperture was positioned on the sphere's bottom surface, coaxial with the LED source. This aperture size was matched to the effective light-emitting area of our composite freeform lens ($\approx$314 mm^2) to eliminate geometric bias in uniformity and efficiency comparisons. And a circular optical baffle (diameter = 15 mm, thickness = 1 mm) was installed along the central axis of the integrating sphere, positioned 30 mm above the exit aperture. This baffle blocks direct light from the LED source to the exit aperture, ensuring that only light scattered by the sphere's inner diffuse surface exits through the aperture—consistent with the working principle of standard integrating spheres for indirect illumination characterization and eliminating direct illumination bias in uniformity/efficiency measurements. Experimental data from three different lighting systems were compared by adjusting the distance from the light source to the target illumination plane and the radius of the target illumination plane. The results indicate that the lighting system designed in this paper improves both the irradiance uniformity and illumination efficiency of the target plane across various scenarios, thus verifying the effectiveness of the combined freeform surface design method. This method is applicable to various fields requiring uniform illumination.

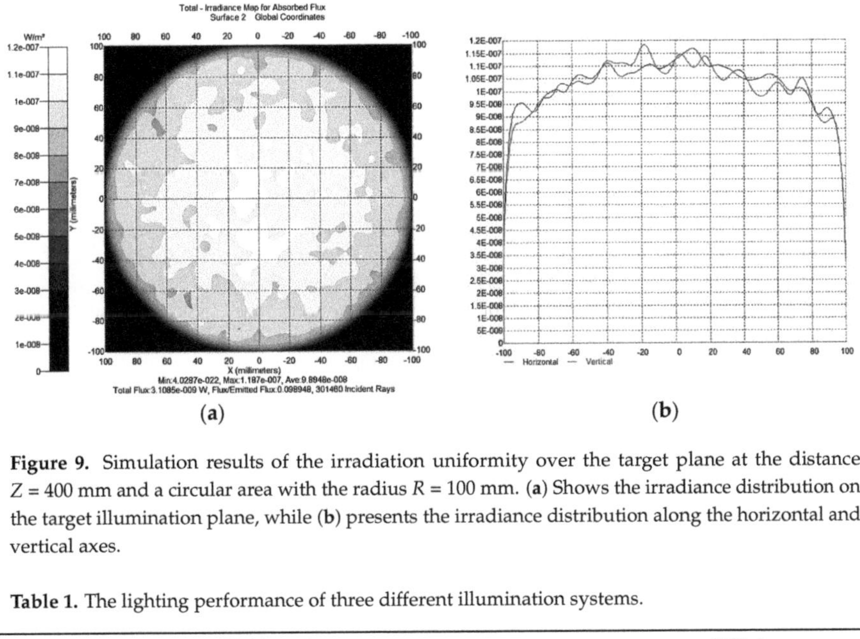

(**a**) (**b**)

Figure 9. Simulation results of the irradiation uniformity over the target plane at the distance $Z = 400$ mm and a circular area with the radius $R = 100$ mm. (**a**) Shows the irradiance distribution on the target illumination plane, while (**b**) presents the irradiance distribution along the horizontal and vertical axes.

Table 1. The lighting performance of three different illumination systems.

Distance	Uniformity			Efficiency		
	Hybrid Diffuse-TIR Freeform Surface	Diffuse-Only Freeform Surface	Integrating Sphere	Hybrid Diffuse-TIR Freeform Surface	Diffuse-Only Freeform Surface	Integrating Sphere
50 mm	84.13%	75.91%	73.13%	41.12%	38.78%	37.13%
100 mm	83.36%	74.83%	72.69%	42.35%	41.31%	38.78%
200 mm	82.87%	73.9%	72.57%	43.37%	41.93%	39.21%
300 mm	82.65%	73.78%	72.01%	43.28%	42.06%	40.17%
400 mm	82.51%	73.51%	71.83%	43.33%	42.11%	40.35%
500 mm	82.46%	73.37%	71.78%	43.39%	42.17%	40.79%

From Table 1, the practical optical efficiency of our hybrid TIR-diffuse system ranges from 41.12 to 43.39%, outperforming diffuse-only (38.78–42.17%) and integrating sphere (37.13–40.79%) systems. And we can be observed that approximately 50% of the total light rays emitted by the LED reach the target plane via the freeform surface designed in this study (i.e., light utilization efficiency), whereas traditional diffuse reflection systems (e.g., spherical diffuse reflectors) only achieve approximately 20% due to Fresnel losses and unregulated scattering—this theoretical advantage is validated by the practical efficiency gap in Table 1. As the illumination distance increases, the illumination efficiency of different illumination systems does not change significantly; however, there is a notable difference in irradiance uniformity. Compared with traditional illumination surfaces, the combined diffuse reflection freeform surface system exhibits superior illumination performance. Figure 10 illustrates the performance variation trends of different lighting systems on the target plane under various conditions.

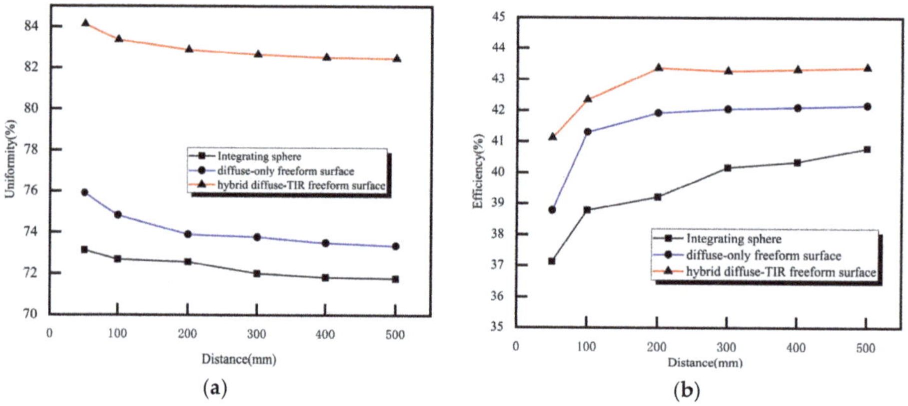

Figure 10. (**a**) Trend chart of irradiance uniformity at different illumination distances with a constant radius of the target illumination plane, (**b**) Trend chart of irradiance efficiency at different illumination distances with a constant radius of the target illumination plane.

Figure 10 illustrates the trends of irradiance uniformity and efficiency under different target illumination planes and illumination distances. It can be observed that the combined diffuse reflection freeform surface designed in this paper outperforms other traditional illumination systems in terms of both irradiance uniformity and illumination efficiency. Thus, the combined diffuse reflection freeform surface illumination system designed in this paper can be considered a feasible approach to achieve high irradiance uniformity and illumination efficiency.

4. Discussion

Due to the insufficient performance of LED lenses with a single freeform surface, the design method of double freeform surface lenses has attracted significant attention. However, a dual freeform surface cannot be uniquely determined without introducing additional constraints. Generally, the luminous flux efficiency and illuminance uniformity of dual freeform surface lenses are significantly superior to those of single freeform surface lenses. Compared with traditional diffuse reflection freeform surfaces, although Fresnel loss is minimized, luminous flux efficiency cannot be improved. Therefore, based on the emission characteristics of a single LED Lambertian source, this paper designs a combination of total internal reflection and diffuse reflection freeform surfaces. A key innovation of this work is the synergistic integration of TIR and diffuse reflection—a combination unattainable with transmissive designs. TIR relies on total internal reflection at the lens-air interface, which requires a continuous, low-loss optical path. Transmissive diffuse media disrupt this path by introducing refractive index inhomogeneities, preventing TIR from occurring (as scattered light cannot maintain the critical angle for total internal reflection). In our reflective design, the diffuse surface acts as a "secondary receiver" for TIR-reflected light (Path 1 in Figure 2): TIR collimates diverging LED rays and directs them to the reflective surface, which then scatters the light uniformly onto the target plane. The integration of TIR and diffuse reflection enhances optical efficiency by approximately 1–4% compared to conventional systems: at the same target distance (e.g., 500 mm), our hybrid system achieves 43.39% efficiency—approximately 1.22% higher than the diffuse-only system (42.17%) and approximately 2.60% higher than the integrating sphere (40.79%); at 50 mm, the hybrid system (41.12%) outperforms the diffuse-only (38.78%) and integrating sphere (37.13%) by approximately 2.34% and approximately 3.99%, respectively (efficiency comparisons in this study strictly follow "same-distance" principles, eliminating misleading cross-distance benchmarks).

It employs a total internal reflection structure to reduce the deflection angles of both incident and outgoing light rays, with the minimum Fresnel loss as a constraint condition. The 3-D space solution problem is compressed into several 2-D planes, and mathematical models for the total internal reflection freeform surface and diffuse reflection freeform surface are established according to Snell's law. Subsequently, an iterative algorithm and the Runge–Kutta algorithm are used to solve the cross-sections of the two freeform surfaces in multiple directions (i.e., 2-D curves). These cross-sections are then closed and laid out using the 3-D modeling software SolidWorks to obtain the combined freeform surface lens. Simulation experiments have validated that its illumination uniformity and efficiency are superior to those of traditional diffuse reflection freeform surfaces. However, the method proposed in this study still has room for improvement, particularly as the lighting efficiency remains at a relatively low level. Feedback optimization design is also a critical component in LED lighting design. This paper only improves the accuracy of the surface morphology by increasing the number of cross-sections, without introducing the irradiance distribution of the illuminated surface as a feedback parameter for surface shape optimization. Therefore, in subsequent work, after experimental analysis, the irradiance of the target surface obtained from simulation experiments can be substituted into the original mathematical model. The model can then be adjusted to optimize the surface shape of the resulting freeform surface, thereby further enhancing the irradiance uniformity and illumination efficiency of the target surface.

In addition, this paper selected a single LED for investigation, while excluding the consideration of multiple LED arrays from the research scope. Regarding extending to multiple LEDs, we will consider the critical angle constraint for TIR in subsequent research. For LEDs in a multi-LED array, rays from edge LEDs may initially propagate at angles below θ_c (failing to meet TIR conditions). To address this, we propose two feasible modifications, which will be added to the Discussion section of the revised manuscript: (i) Spatial arrangement optimization: Position each LED such that its incident angle on the TIR surface falls within $[\theta_c, 90^\circ]$. This is achievable by adjusting the array's pitch and distance from the TIR surface (e.g., a hexagonal array with LED centers spaced 1.2× the TIR surface's local curvature radius). (ii) Hybrid optical path: For rays below θ_c, integrate a micro-prism array on the TIR surface to redirect these rays to meet TIR conditions (micro-prisms adjust the incident angle by ~5–15°, ensuring $\theta \geq \theta_c$.

5. Conclusions

This paper develops an illumination system combining total internal reflection and diffuse reflection freeform surfaces based on a multiple diffuse reflection freeform surface model. Experimental results validate its effectiveness and feasibility. The composite freeform surface illumination model designed in this paper exhibits high irradiance uniformity and illumination efficiency, which is significantly superior to other traditional diffuse reflection illumination systems. Furthermore, this method has a certain degree of universality: it only requires optimizing and adjusting the target plane parameters to meet the lighting requirements of other fields, thereby effectively expanding the application prospects of diffuse indirect lighting in industrial production and daily life.

Author Contributions: Conceptualization, X.X., J.R. and Z.Z.; methodology, X.X., X.L. and Y.P.; software, X.X. and Y.P.; validation, X.X., J.R. and Z.Z.; formal analysis, X.X., J.R. and Y.P.; investigation, J.R., X.L. and Z.Z.; resources, J.R., X.L. and Y.P.; data curation, X.X. and Z.Z.; writing—original draft preparation, X.X.; writing—review and editing, X.X., J.R. and X.L.; visualization, X.X. and Z.Z.; supervision, J.R. and X.L.; project administration, X.X., J.R. and Z.Z.; funding acquisition, X.X., J.R. and Z.Z. All authors have read and agreed to the published version of the manuscript.

Funding: This research was funded by Science and Technology Project of Jiangxi Provincial Department of Education (GJJ2403802), Jiangxi Provincial Natural Science Foundation (20242BAB25104, 20242BAB25057), Key R&D Program Project of Jiangxi Province (20232BBE50012), and Education Department Science and Technology Plan Project-Youth Science Fund Project (GJJ2403806).

Institutional Review Board Statement: Not applicable.

Informed Consent Statement: Not applicable.

Data Availability Statement: The data underlying the results presented in this paper are not publicly available at this time, but may be obtained from the authors upon reasonable request.

Conflicts of Interest: The authors declare no conflicts of interest.

References

1. Chiang, C.W.; Hsu, Y.K.; Pan, J.W. Design and demonstration of high efficiency anti-glare LED luminaires for indoor lighting. *Opt. Exp.* **2015**, *23*, 12–23. [CrossRef] [PubMed]
2. Jiang, J.; To, S.; Lee, W.; Cheung, B. Optical design of a freeform TIR lens for LED streetlight. *Optik* **2010**, *121*, 1761–1765. [CrossRef]
3. Wang, J.; Liang, Y.; Xu, M. Design of a See-Through Head-Mounted Display with a Freeform Surface. *J. Opt. Soc. Korea* **2015**, *19*, 614–618. [CrossRef]
4. Wang, X.; Li, X.; Zhang, Q.; Wei, W.; Liu, E. Efficient Testing Light Path for Aspherical Surfaces Based on Secondary Imaging. *Photonics* **2025**, *12*, 485. [CrossRef]
5. Pan, J.-W.; Su, Y.-C.; Chen, Y.-S. Secondary optical element design for intracorporeal LED illumination system. *Opt. Lett.* **2014**, *39*, 224–227. [CrossRef]
6. Cai, X.; Quan, L.; Wu, J.; He, Y. Night-time disability glare of constant-light LED traffic monitoring fill light. *Light. Res. Technol.* **2021**, *53*, 595–608. [CrossRef]
7. Ren, C.; Meng, Q. Design Method for Freeform Off-Axis Three-Mirror Anastigmat Optical Systems with a Large Field of View and Low Error Sensitivity. *Photonics* **2024**, *11*, 211. [CrossRef]
8. Xu, S.; Peng, X.; Song, C. Adaptive Freeform Optics Design and Multi-Objective Genetic Optimization for Energy-Efficient Automotive LED Headlights. *Photonics* **2025**, *12*, 388. [CrossRef]
9. Li, Y.; Feng, Z.; Cheng, D.; Wang, Y. Designing double freeform surfaces for large ray bending irradiance tailoring of extended LED sources. *Opt. Express* **2021**, *29*, 13469–13485. [CrossRef] [PubMed]
10. Huang, J.; Jiang, S.; Yang, H.; Czuma, R.; Yang, Y.; Kozel, F.A.; Jiang, H. Portable diffuse optical tomography for three-dimensional functional neuroimaging in the hospital. *Photonics* **2024**, *11*, 238. [CrossRef]
11. Liu, X.; Zhu, J. Design method of freeform off-axis multi-mirror optical systems. *Photonics* **2022**, *9*, 534. [CrossRef]
12. Liu, B.; Jin, X.; Zhu, Z. Uniform illumination design based on LED array and diffuse reflection freeform surface. *J. Appl. Opt.* **2014**, *35*, 598–602.

13. Tsuei, C.-H.; Sun, W.-S. Momentary adjustment means for simulating the sunlight color temperature, hues and brightness with RGB LEDs in indoor lighting. *Phys. Procedia* **2011**, *19*, 239–243. [CrossRef]
14. Zhu, Z.-M.; Qu, X.-H.; Jia, G.-X.; Ouyang, J.-F. Uniform Illumination Design by Configuration of LED Array and Diffuse Reflection Surface for Color Vision Application. *J. Disp. Technol.* **2011**, *7*, 84–89. [CrossRef]
15. Zhu, Z.-M.; Liu, H.; Chen, S.-M. The Design of Diffuse Reflective Free-Form Surface for Indirect Illumination with High Efficiency and Uniformity. *IEEE Photonics-J.* **2015**, *7*, 1–10. [CrossRef]
16. Zhu, Z.-M.; Sun, X.; Peng, B. The Design of Diffuse Reflective Off-Axis Surface for Noncircular LED Arrays. *IEEE Photonics-J.* **2016**, *9*, 1600210. [CrossRef]
17. Lan, D.; Wei, W. TIR Lens-based optical system design for dynamic star simulator illumination. *Illum. Eng. J.* **2023**, *34*, 22–28.
18. Zhu, Z.; Xu, X.; Sun, X.; Wang, X. Establishment and solution for a mathematical model of multiple diffuse reflection. *IEEE Photonics-J.* **2020**, *12*, 6500917. [CrossRef]

photonics

MDPI

Article

Investigation of the Damage Characteristics and Mechanisms in Silicon Carbide Crystals Induced by Nanosecond Pulsed Lasers at the Fundamental Frequency

Penghao Xu [1,2], Erxi Wang [1], Teng Wang [1], Chong Shan [1,3,*], Xiaohui Zhao [1], Huamin Kou [3], Dapeng Jiang [3], Qinghui Wu [3], Zhan Sui [1] and Yanqi Gao [1]

1 Shanghai Institute of Laser Plasma, China Academy of Engineering Physics, 1129 Chenjiashan Road, Shanghai 201800, China
2 Troops 91007, Chinese People's Liberation Army, Shanghai 201800, China
3 State Key Laboratory of Functional Crystals and Devices, Shanghai Institute of Ceramics, Chinese Academy of Sciences, Shanghai 201899, China
* Correspondence: shanchong@mail.sic.ac.cn

Abstract

Silicon carbide (SiC) single crystals are extensively utilized in various fields due to their exceptional properties, such as a wide bandgap and a high breakdown threshold. Nevertheless, the intrinsic high hardness of SiC creates significant challenges for contact machining. This study investigates the surface damage characteristics and underlying mechanisms involved in processing both high-purity silicon carbide (HP-SiC) and nitrogen-doped silicon carbide (N-SiC) crystals using fundamental-frequency nanosecond pulsed lasers. This study establishes a laser-induced damage threshold (LIDT) testing platform and employs the internationally standardized 1-ON-1 test method to evaluate the damage characteristics of HP-SiC and N-SiC crystals under single-pulse laser irradiation. Experimental results indicate that N-SiC crystals exhibit superior absorption characteristics and a lower LIDT compared with HP-SiC crystals. Subsequently, a defect analysis model was established to conduct a theoretical examination of defect information across various types of SiC. Under fundamental-frequency nanosecond pulsed laser irradiation, N-SiC crystals demonstrate a lower average damage threshold and a broader defect damage threshold distribution than their HP-SiC counterparts. By integrating multi-dimensional analytical methods—including photothermal weak absorption mechanisms and damage morphology analysis—the underlying damage mechanisms of the distinct SiC forms were comprehensively elucidated. Moreover, although N-SiC crystals show weaker photothermal absorption properties, they exhibit more pronounced absorption and damage response processes. These factors collectively account for the different laser damage resistances observed in the two types of silicon carbide crystals, implying that distinct processing methodologies should be employed for nanosecond pulsed laser treatment of different SiC crystals. This paper elucidates the damage characteristics of various SiC materials induced by near-infrared nanosecond pulsed lasers and explores their underlying physical mechanisms. Additionally, it provides reliable data and a comprehensive mechanistic explanation for the efficient removal of these materials in practical applications.

Keywords: silicon carbide single crystals; laser-induced damage threshold; defect analysis model; photothermal weak absorption

1. Introduction

Silicon carbide (SiC) crystals, recognized as being among the third-generation semiconductor materials, exhibit exceptional properties such as a wide bandgap, high thermal conductivity, and a high breakdown threshold [1–7]. These attributes render them indispensable for applications spanning aerospace, new energy vehicles, controlled fusion diagnostics, and various sectors within the electronic and information industries [8–14]. SiC single crystals are increasingly utilized across diverse fields, driving a growing demand for efficient and precise subtractive machining techniques such as cutting and drilling. However, owing to their exceptional hardness and brittle nature—with hardness second only to diamond—these crystals present significant challenges for precision machining. Pulsed lasers exhibit characteristics such as short pulse durations and high-power densities. Unlike traditional contact machining methods, such as diamond wire cutting, pulsed laser processing offers advantages in efficiency, precision, and minimized material loss. Consequently, these attributes make pulsed laser machining a highly promising approach for material removal in the processing of SiC crystals.

There have been numerous reports on the non-contact precision machining of SiC crystal using pulsed lasers [15–22]. Wu et al. employed femtosecond pulsed lasers to investigate the removal mechanisms of SiC crystal in various environments, including air, water, and hydrofluoric acid solutions [20]. Their findings demonstrate that the processing environment significantly influences the laser ablation threshold. Tomita et al. investigated the formation of micro- and nanostructures on 4H-SiC crystal surfaces via ultrashort laser pulse irradiation. They found that introducing artificially prepared scratches with specific orientations effectively reduced the incident laser energy density required to generate high-spatial-frequency structures, while exerting minimal influence on the threshold for inducing low-spatial-frequency features [22]. Previous studies have primarily focused on investigating the characteristics and mechanisms underlying the ultra-short pulse laser processing of SiC crystals. In contrast, a systematic exploration of the damage characteristics and mechanisms induced by nanosecond pulse lasers in these crystals remains to be conducted. Compared with femtosecond and picosecond pulse lasers, nanosecond pulse lasers introduce thermal effects during processing, which preclude cold processing conditions and consequently reduce machining precision. Nevertheless, nanosecond pulse lasers can achieve higher power outputs, presenting significant potential for applications involving large-scale material removal in SiC crystals. Therefore, further research into the associated material damage mechanisms is urgently needed. This study diverges from previous investigations on the mechanisms of ultra-short pulse laser precision machining of silicon carbide by addressing the efficiency and cost challenges associated with high-efficiency cutting of large-diameter silicon carbide crystals for industrial applications. It analyzes the material behavior during nanosecond pulse laser processing, demonstrating its suitability for these challenging scenarios. Moreover, the study examines the damage characteristics and mechanisms induced by laser irradiation in silicon carbide crystals using a multidimensional approach that incorporates weak photothermal absorption analysis, defect modeling, and damage morphology assessment. These findings provide comprehensive theoretical and experimental support for advancing the efficient nanosecond pulsed laser processing of silicon carbide crystals in future industrial applications.

This study investigates the damage characteristics and underlying mechanisms in high-purity silicon carbide (HP-SiC) and nitrogen-doped silicon carbide (N-SiC) crystals induced by fundamental-frequency nanosecond pulsed lasers. Initially, using the 1-ON-1 laser-induced damage threshold (LIDT) testing method, we measured the LIDT on the surfaces of HP-SiC and N-SiC crystals. Subsequently, by incorporating defect analysis models, we obtained critical defect information—including defect density, average damage threshold, and damage threshold distribution—for the various types of SiC crystal. Meanwhile, a photothermal weak absorption testing platform was developed to characterize the absorption properties of fundamental-frequency laser pulses on various SiC crystal surfaces. This platform facilitated the analysis of absorption characteristics associated with distinct crystal defects and verified the defect information obtained from a defect analysis model. Furthermore, differences in damage morphologies between HP-SiC and N-SiC were examined to elucidate the underlying damage mechanisms. This study systematically investigates the physical mechanisms responsible for damage induced by fundamental-frequency nanosecond pulsed lasers on SiC crystal surfaces and provides critical data to support the precise control of SiC crystal processing using these lasers.

2. Experimental Details

2.1. Test Sample

In this study, HP-SiC and N-SiC crystals were utilized as subjects. To minimize the influence of processing factors when comparing the damage characteristics and mechanisms of these two SiC types, both crystal surfaces underwent precision polishing. This process achieved a surface roughness of less than 0.1 nm and a Total Thickness Variation (TTV) below 1.6 μm, thereby meeting wafer-grade standards. Due to the distinct properties and functional applications of the two SiC crystals (as shown in Table 1), HP-SiC—with its exceptional thermal conductivity, hardness, and stability—serves as an ideal material for lens substrates and crystal heat sinks. Consequently, it is processed into circular samples (Φ 17 mm × 1 mm) using water-guided laser cutting. In contrast, conductive N-SiC, primarily utilized in semiconductor chip manufacturing, is processed into square samples (10 mm × 10 mm × 1 mm) via grinding wheel machining. Finally, to objectively assess the nanosecond laser-induced damage characteristics of various SiC crystals, all laser irradiation experiments were conducted on the C-plane of both HP-SiC and N-SiC [23].

Table 1. The physical parameter of HP-SiC and N-SiC crystals.

Parameters	HP-SiC	N-SiC
Crystal structure	4H	4H
Crystal surface	$(000\bar{1})$ C-plane	$(000\bar{1})$ C-plane
Total Thickness Variation (TTV)	<1.6 μm	<1.6 μm
Density (25 °C)	3.21 g/cm^3	3.21 g/cm^3
Resistivity	1010 Ω·cm	22 mΩ·cm
Near-field spot size	20 μm	20 μm

2.2. Laser-Induced Damage Testing Platform

This study establishes a damage threshold testing platform for fundamental-frequency nanosecond pulsed lasers, as depicted in Figure 1. The Nd:YAG pulsed laser employed in the investigation emits a nanosecond pulse at a central wavelength of 1064 nm, with a pulsed duration of 7.6 ns (full width at half maximum) and a near-field spot size of 6.8 mm (at 1/e) under S-polarization. After passing through a laser energy attenuation system composed of a half-wave plate and a polarizer, the laser beam is directed toward a beam splitter. It splits the beam into two distinct paths: one leading to an energy meter and the other to a beam quality analyzer. This configuration enables real-time monitoring of the energy levels and near-field spot characteristics during the LIDT testing process. The incident laser beam passes through the beam splitter in the transmission direction and is subsequently focused by a lens with a 300 mm focal length, ultimately irradiating the sample surface with a focal-spot area of 0.04 mm^2. A CCD camera equipped with a 20× magnification system is employed to monitor, in real time, any damage to the incident surface of the SiC crystal. Residual laser energy transmitted through the test sample is absorbed by an absorption baffle. The specific laser parameters for the LIDT test are shown in Table 2.

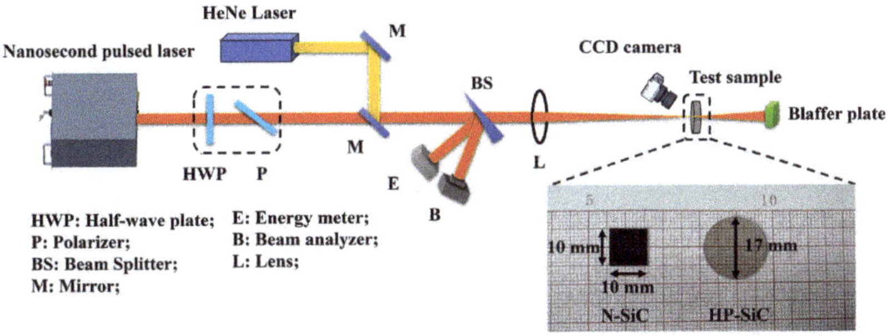

Figure 1. Schematic diagram of the LIDT test system.

Table 2. The specific laser parameters for the LIDT test.

LIDT Test	Optical Parameter
Wavelength	1064 nm
Duration	7.6 ns
Near-field spot size	6.8 mm
Polarization	S
Focused length	300 mm
Focal-spot area	0.04 mm^2

The laser damage threshold test method used in this study is the 1-ON-1 test method, as specified in the international standard (ISO 21254) [24]. This method, also known as the single-pulse laser damage threshold test, involves irradiating each position on the test specimen with a single laser pulse; regardless of whether damage occurs, the irradiation then moves to the next area. A CCD camera is employed to record any irreversible structural changes on the specimen following laser irradiation, thereby determining whether the laser exposure induces material damage. Single pulsed laser irradiation was applied at multiple positions on the test samples using a constant incident laser energy density. The probability of laser-induced damage was quantified as the ratio of observed damage events to the total number of irradiation trials, yielding a damage probability. By recording these probabilities across a range of energy density levels and performing linear regression, the incident laser energy density corresponding to zero damage probability was determined. This value was then established as the laser damage threshold for the material.

To compare the nanosecond pulsed laser damage thresholds of the two materials, it is essential to carefully select the laser fluence steps and determine the number of irradiations per fluence step. Although the sample sizes of the HP-SiC and N-SiC crystals differed, a consistent approach was adopted by selecting twelve fluence steps for each crystal, with ten irradiations conducted at each step, thereby ensuring an objective assessment of their nanosecond laser damage resistance. This methodology not only enables an accurate comparison of the laser damage thresholds but also facilitates the acquisition of more precise defect information.

2.3. Photothermal Weak Absorption Test Apparatus

Photothermal weak absorption leverages the thermal effects induced in materials during the heating process from pump laser interaction, which in turn leads to measurable phase shifts in the probe light. These phase shifts allow for theoretical quantification of the material's photothermal weak absorption response to the pump laser. In this study, a pump laser with a central wavelength of 1064 nm and an incident power of 2.5 W is employed. A chopper maintains the laser frequency at 200 Hz, while a 1064 nm half-wave plate in combination with a polarizer enables precise adjustment of the pump laser's incident power. Subsequently, a focusing lens directs the pump laser onto the test sample surface, achieving a spot diameter of 20 μm. The probe light is generated by a He-Ne laser with a central wavelength of 632.8 nm. Precise adjustment of the incident laser power is achieved using an attenuator composed of a half-wave plate and a polarizer, both optimized for 632.8 nm. Through coordinated adjustments of mirrors and focusing lenses, the probe light spot is made comparable in size to the pump light spot, facilitating overlapping irradiation on the sample surface, as illustrated in Figure 2. The sample is translated in increments of 0.05 mm, with each test covering an area of 2.5 mm × 2.5 mm. By measuring the photothermal weak absorption characteristics at multiple positions on the sample, the absorption properties of various surface regions are determined. When the SiC crystal surface absorbs the pump laser, localized temperature increases occur. These thermal effects induce phase changes in the detection light, enabling the derivation of the corresponding photothermal weak absorption coefficient through theoretical calculations. The corresponding parameters for the photothermal weak absorption testing system are shown in Table 3.

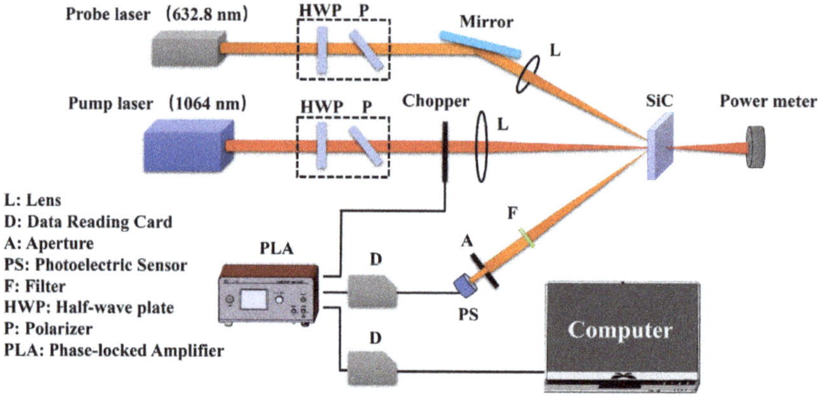

Figure 2. Schematic diagram of the photothermal weak absorption test system.

Table 3. The specific laser parameters for the photothermal weak absorption test.

Photothermal Weak Absorption Test	Pump Light	Probe Light
Wavelength	1064 nm	632.8 nm
Power	2.5 W	0.6 W
Duration	200 Hz	CW
Near-field spot size	20 μm	20 μm

Regarding the photothermal weak absorption testing, the relevant testing procedures were as follows: First, a pulsed pump laser, centered at a 1064 nm wavelength, is used. After passing through an attenuator consisting of a half-wave plate and a polarizer, the pump laser power is modulated. To ensure that sufficient power induces crystal defect absorption and generates the photothermal effect without causing material damage, the pump laser power was set at 2.5 W. Subsequently, the pump laser beam was directed through a chopper operating at 200 Hz, which transmits the modulated signal to a phase-locked amplifier. The beam was then focused by a lens, forming a 20 μm diameter focal spot on the sample surface. Finally, after traversing the sample, the laser power was measured in real-time using a power meter. Subsequently, continuous light emitted by a He-Ne laser with a central wavelength of 632.8 nm was used as the detection beam. After passing through an attenuator—comprising a half-wave plate and polarizer, both optimized for 632.8 nm—the detection beam power is adjusted to 0.6 W, ensuring that it does not interfere with the interaction between the pump light and crystal defects. The incidence direction of the detection light is then modified using a mirror, and a focusing lens concentrates the beam, maintaining identical positions and sizes for both the pump and detection beams on the test sample surface. The detection light reflected from the crystal surface passes through a filter and aperture before being captured by a photodetector. The signal was subsequently transmitted to a card reader and then processed by a phase-locked amplifier, which recorded the phase shift in the detection light induced by the thermo-optical effect during the interaction between the pump laser and the crystal. This measurement enables a theoretical calculation of the material's photothermal weak absorption coefficient corresponding to the thermo-optical effect during the pump laser-SiC crystal interaction. It was worth noticing that the photothermal weak absorption test system described herein constitutes a comparative measurement method. Its principle is based on the generation of a thermal envelope effect when pump light (1064 nm) interacts with crystal defects. Subsequently, as probe light passes through the region exhibiting this thermal envelope, its phase is altered. By recording these phase shifts, one can theoretically calculate the relative weak absorption value at a given position in comparison to defect-free crystal regions. Unlike laser calorimetry, which provides absolute measurements, this method employs the photothermal weak absorption values of calibrated components as reference standards, rendering it only suitable for comparing the weak absorption characteristics of different crystals rather than determining precise photothermal weak absorption values.

3. Experimental Results

Table 4 presents the results of one-on-one laser damage threshold tests for HP-SiC and N-SiC crystals subjected to fundamental-frequency nanosecond pulsed lasers. The test standard employed in this study is the single-pulse laser damage threshold method (1-ON-1 LIDT test method), as defined by ISO 21254 [24]. This method entails irradiating a specimen with a single pulse of laser light at a designated location. Regardless of whether damage occurs, the laser is subsequently repositioned to a new location for further testing. By varying the incident laser fluence across a series of steps and recording the corresponding damage probabilities, the laser fluence at which no damage is observed is determined and designated as the LIDT. This LIDT serves as a quantitative measure of the material's resistance to initial damage. The data indicate that HP-SiC crystals exhibit a higher laser damage threshold than N-SiC crystals. It was worth noting that during the LIDT test, identical laser fluence steps and laser irradiation shots were employed to compare the nanosecond pulsed laser damage thresholds at the fundamental frequency for various SiC crystals, thereby ensuring accurate comparative measurements. To mitigate the effects of disparate crystal growth processes on the evaluation of laser damage resistance, and to objectively analyze the damage mechanisms in different SiC crystals, five HP-SiC and five N-SiC crystals were selected from distinct batches. The presented results correspond to the experimental outcome closest to the average threshold value among the five tests, ensuring both experimental precision and statistical reliability. Additionally, for the photothermal weak absorption evaluation, five distinct locations on each of the five samples were tested, confirming the accuracy of the photothermal weak absorption test in these SiC crystals. This finding provides critical insights for the precision machining of both types of SiC crystals using fundamental-frequency nanosecond pulsed lasers. For instance, when employing these lasers for the precision cutting and wafering of large-diameter SiC crystals, it is essential to strictly control the incident laser energy density for N-SiC crystals near their damage threshold to minimize processing losses. In contrast, for HP-SiC crystals, increasing the incident laser energy density appropriately can enhance processing efficiency and productivity.

Table 4. Laser-induced damage thresholds of HP-SiC crystal and N-SiC crystal.

	HP-SiC (J/cm^2)	N-SiC (J/cm^2)
LIDT	12.04	8.06

Meanwhile, the characteristics and mechanisms underlying multi-pulse cumulative irradiation of SiC crystals pose significant challenges for processing these materials using nanosecond pulsed lasers. This topic remains a critical focus of our ongoing research. In this study, we conducted a series of investigations addressing these challenges. Initially, using the S-ON-1 laser-induced damage threshold testing method (frequency = 5 Hz, S values = 5, 10, 30, 50), we determined the multi-pulse laser damage thresholds for two types of SiC crystals. The test results are summarized in Table 5. Experimental results indicate that under identical repetition rates and multi-pulse laser irradiation conditions, both HP-SiC and N-SiC crystals exhibited a reduction in LIDT, with the decrease becoming more pronounced as the number of pulses increased. Notably, the decline in multi-pulse LIDT was more significant for N-SiC crystals than for HP-SiC crystals. This observation suggests that, under multi-pulse laser irradiation, the cumulative effects enhance defect-mediated absorption in N-SiC crystals to a greater extent than in HP-SiC crystals. The more pronounced reduction in LIDT for N-SiC might be attributed to the diverse lattice defects introduced during its doping process. Consequently, when employing fundamental-frequency nanosecond pulsed lasers of identical fluence for multi-pulse irradiation of different SiC crystals, rigorous control over the number of pulses and repetition rate is essential for N-SiC crystals to prevent excessive ablation.

Table 5. The LIDTs induced in HP-SiC and N-SiC crystals under different pulse number irradiation at the same repetition rate (Frequency = 5 Hz).

	HP-SiC (J/cm^2)	N-SiC (J/cm^2)
S = 5	10.62	6.17
S = 10	8.89	4.36
S = 30	6.38	2.63
S = 50	5.01	1.02

Moreover, variations in multi-pulse laser damage thresholds across diverse laser fluence gradients have also been examined. We employed the R-ON-1 laser damage threshold testing method, whereby each location is subjected to irradiation with progressively increasing energy densities in increments of 0.5 J/cm^2, as detailed in Table 6. The results demonstrate that the R-ON-1 laser damage thresholds for both types of SiC crystals exceed their respective 1-ON-1 damage thresholds, indicating an effective laser conditioning effect. Notably, the threshold enhancement is more pronounced in N-SiC crystals compared to HP-SiC crystals, suggesting that the laser conditioning more effectively mitigates threshold-lowering lattice defects introduced during the doping process of N-SiC crystals. The underlying physical mechanism will be investigated in subsequent studies.

Table 6. The LIDTs for both HP-SiC and N-SiC crystals were determined using the R-ON-1 laser damage threshold test method.

	HP-SiC (J/cm^2)	N-SiC (J/cm^2)
LIDT (R-ON-1)	14.65	12.81

4. Discussion

4.1. *Spectral Absorption Testing*

To elucidate the physical mechanisms underlying the differences in laser damage thresholds observed in two types of SiC crystals, this paper examines four key aspects: spectral absorption, defect characterization, weak photothermal absorption properties, and damage morphology analysis. Firstly, this study conducted absorption spectroscopy measurements on two types of SiC crystals using a Cary 5000 spectrophotometer, Santa Clara, CA, USA (see Figure 3). The spectral range spanned from 200 nm to 1500 nm. Notably, at a wavelength of 1064 nm in the near-infrared region, HP-SiC exhibited a transmittance of 66%, whereas N-SiC showed only 18%. These findings indicate that N-SiC crystals possess significantly higher absorption, which is a primary factor contributing to their reduced damage threshold.

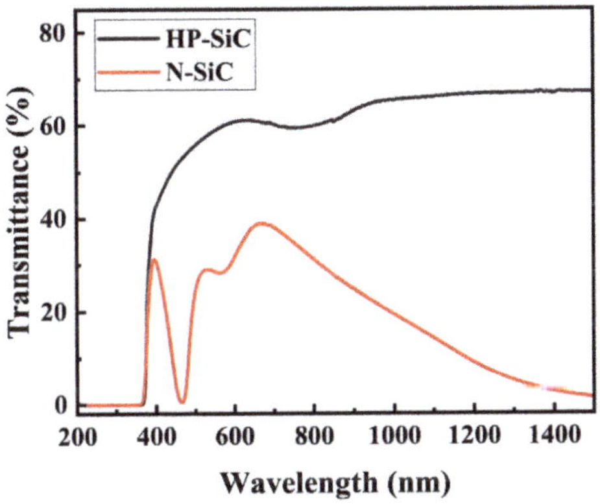

Figure 3. Absorption spectrum test results for two types of silicon carbide crystals.

However, the damage threshold did not exhibit the substantial variation observed in the absorption spectrum. This observation indicates that the damage properties of SiC crystals induced by fundamental-frequency nanosecond lasers are not solely determined by the material's spectral absorption characteristics; rather, additional physical mechanisms are involved. Subsequently, this paper presents separate analyses from the perspectives of weak photothermal absorption, defect characterization, and damage morphology and processes.

4.2. *Defect Precursor Information*

The mechanisms underlying laser-induced damage differ between nanosecond and ultrafast pulsed lasers, with material defect information playing a crucial role in determining damage characteristics. Silicon carbide crystals exhibit a diverse range of surface defects—including both intrinsic crystal flaws and those introduced during processing—which display varying capacities for laser absorption and corresponding resistance to laser damage. Moreover, these defects are randomly distributed across the crystal surface. Consequently, acquiring detailed information about these defects is essential for accurately analyzing the damage characteristics of HP-SiC and N-SiC crystals.

This study employs the statistical model proposed by Krol et al. [25,26], which assumes that defect damage thresholds follow a Gaussian distribution, in conjunction with the results of the 1-ON-1 laser damage threshold tests described above, to analyze defect characteristics in two types of SiC crystals. The model postulates that both the mean defect damage threshold (T_0) and the defect damage threshold distribution (ΔT) within the test specimen adhere to a normal distribution function. Consequently, the corresponding defect function can be expressed as:

$$g(T) = \frac{2d}{\Delta T\sqrt{2\pi}} exp\left[-\frac{1}{2}\left(\frac{T - T_0}{\frac{\Delta T}{2}}\right)^2\right] \quad (1)$$

$$\mathrm{d} = \int_0^{\infty} g(T)dT. \quad (2)$$

During damage testing, the total number of defects within a specified spot area is integrated to yield the defect density (d). In 1-ON-1 laser damage threshold testing, this result represents the probability of damage corresponding to different incident laser energy densities. Consequently, it is necessary to combine the aforementioned defect analysis model with the Poisson distribution to establish a connection between the defect statistical model and the damage probability. Within the equivalent spot area, $S_{eff} = \pi\omega_x \times \omega_y/2$, the region where the laser energy density exceeds the threshold ($F > T$) is denoted as $S_T(F)$:

$$S_T(F) = S_{eff}\ln\left(\frac{F}{T}\right). \quad (3)$$

$$N(F) = \int_0^F g(T)S_T(F)dT. \quad (4)$$

In this equation, $N(F)$ represents the number of defects within the laser-illuminated region where the energy density (F) exceeds the defect threshold (F) necessary to induce damage. Consequently, the damage probability, denoted as $P(F)$, is expressed using the Poisson formula:

$$P(F) = 1 - exp[1 - N(F)]. \quad (5)$$

As a result, the defect information corresponding to HP-SiC and N-SiC crystals can be obtained, as illustrated in Figure 4 and detailed in Table 7. Notably, the N-SiC crystals exhibit a lower average damage threshold (T_0) than the HP-SiC crystals, indicating that under identical energy density absorption conditions, the surface defects in N-SiC crystals are more prone to laser-induced damage. This disparity is the primary factor leading to the difference in overall single-pulse laser damage thresholds between the two types of SiC crystals. Meanwhile, N-SiC crystals exhibit a broader distribution of defect damage thresholds (ΔT) relative to HP-SiC crystals, suggesting that they harbor a more diverse array of defects. These findings will be corroborated through subsequent analyses of their photothermal weak absorption characteristics.

It is noteworthy that N-SiC crystals exhibit a broader defect damage threshold range while maintaining a lower defect density compared to their HP-SiC counterparts. It was mainly caused by the combined influence of the LIDT test outcomes and defect analysis models. In the LIDT testing, the laser fluence for HP-SiC crystals spans approximately 12.5 $\mathrm{J/cm^2}$ to 17.5 $\mathrm{J/cm^2}$ over the 0% to 100% damage probability spectrum, whereas for N-SiC crystals, the range is approximately 7.5 $\mathrm{J/cm^2}$ to 17 $\mathrm{J/cm^2}$. This observation indicates that the defect damage energy density range for N-SiC crystals is broader than that for HP-SiC crystals. In the defect analysis model presented, defect data simulation involves fitting the defect information within the portion of the incident spot where the laser fluence exceeds the defect damage threshold. Consequently, for HP-SiC crystals, the narrower defect damage fluence range signifies that a smaller area of the incident spot experiences defects spanning the full damage probability spectrum. As a result, the defect fitting procedure yields a higher defect density for HP-SiC crystals compared to N-SiC crystals.

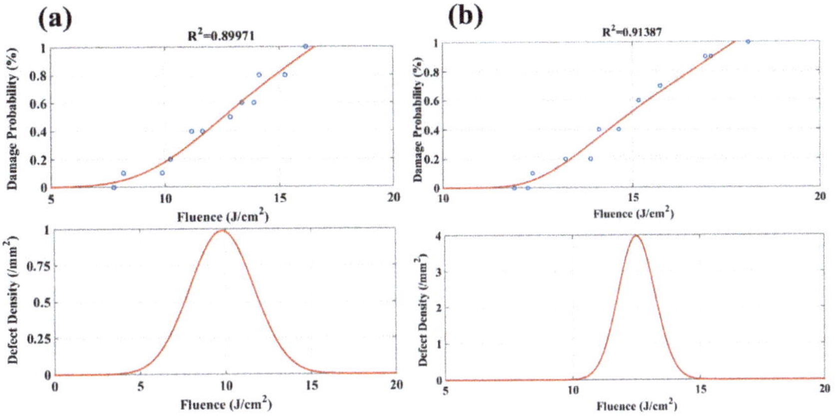

Figure 4. Simulations of two variants of silicon carbide crystals reveal distinct defect characteristics (the blue circles: the LIDT test results; the red lines: the simulation results): (**a**) N-SiC; (**b**) HP-SiC.

Table 7. The defect information of different SiC crystal.

	d (/mm^2)	T_0 (J/cm^2)	ΔT (J/cm^2)
HP-SiC	7.19	12.52	1.44
N-SiC	4.63	9.8	3.74

4.3. Photothermal Weak Absorption Characteristics

Nanosecond pulsed laser-induced material damage is primarily governed by thermal mechanisms arising from the defect absorption of laser radiation. The inherent weak photothermal absorption serves as a critical indicator for characterizing defect distribution and absorption capacity. This approach addresses the "thermal envelope effect" produced during pump laser–material interactions by employing theoretical calculations based on phase changes observed in the recorded probe light, thereby determining the spatial variations in photothermal absorption properties. The experimental setup is depicted in Figure 2.

In this study, to objectively compare the photothermal weak absorption characteristics of HP-SiC and N-SiC crystals, photothermal weak absorption tests were performed at five distinct positions on the surfaces of both SiC types. The test area measured 2.5 mm × 2.5 mm with a step length of 0.05 mm. The results, shown in Figure 5 and summarized in Table 8, indicate that the average photothermal weak absorption value of the HP-SiC crystal (19.68 ppm) is lower than that of the N-SiC crystal (34.23 ppm). This observation suggests that surface defects in N-SiC crystals exhibit stronger laser absorption characteristics than those in HP-SiC crystals, which aligns with earlier findings indicating that N-SiC crystals have a lower average damage threshold (T_0). Moreover, the photothermal weak absorption dynamic range of N-SiC crystals, with a maximum value of 230.17 ppm, exceeds that of HP-SiC crystals, which reached 148.63 ppm. These results further corroborate the defect analysis, indicating that the defect damage threshold range (ΔT) for N-SiC crystals is greater than that for HP-SiC crystals, thereby confirming the accuracy of the defect information obtained for each SiC crystal. Therefore, different

SiC crystals exhibit distinct photothermal weak absorption characteristics, which are the primary factors responsible for variations in their damage thresholds.

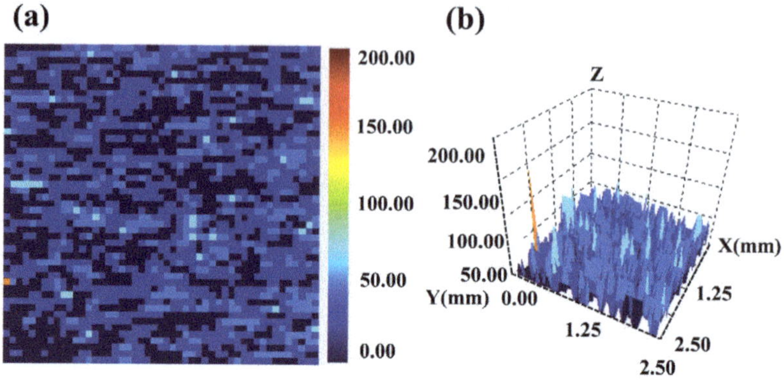

Figure 5. Test results of photothermal weak absorption. (**a**) Two-dimensional imaging result; (**b**) Three-dimensional imaging result.

Table 8. Photothermal weak absorption of HP-SiC crystal and N-SiC crystal.

	HP-SiC	N-SiC
Average value (ppm)	19.68	34.23
Maximum value (ppm)	148.63	230.17

4.4. Damage Morphology Analysis

The preceding discussion analyzed the mechanisms underlying the distinct damage characteristics of the two SiC types by examining spectral absorption, defect analysis, and photothermal weak absorption properties. In this section, we investigate the variations in damage behavior by analyzing the morphologies induced in different SiC crystals by fundamental-frequency nanosecond pulsed lasers. Both HP-SiC and N-SiC crystals were subjected to an identical incident laser energy density of 31.37 J/cm^2 to induce surface damage, and the resulting damage morphologies were documented using Scanning Electron Microscopy (SEM) and Energy Dispersive Spectrometer (EDS), as illustrated in Figure 6.

Firstly, the damage morphology demonstrates that both HP-SiC and N-SiC crystals exhibit ablation under nanosecond pulsed laser irradiation, indicating that thermal effects remain the dominant damage mechanism in both types of SiC at the fundamental frequency. It was worth noticed that the laser damage testing platform employs a spot with an area of 0.04 mm^2, corresponding to a diameter of approximately 120 μm. Consequently, when comparing the damage morphology of different crystals, the damage dimensions of the N-SiC crystal was comparable to the spot size. Moreover, we exposed both SiC crystals to an identical incident laser fluence (31.37 J/cm^2) to induce initial surface damage at five distinct positions. SEM was then employed to record and compare the dimensions and depths of the damage, with the results summarized in Table 9. It could be observed that for the identical incident laser fluence, the damage depths of the two silicon carbide crystals remained essentially consistent, but the surface damage area of N-SiC is consistently larger than that of HP-SiC. This observation further corroborates the previously noted differences in photothermal absorption characteristics between the two materials, with N-SiC exhibiting higher photothermal absorption. Consequently, under equivalent irradiation conditions, N-SiC absorbs more laser energy, resulting in a larger ablation damage area.

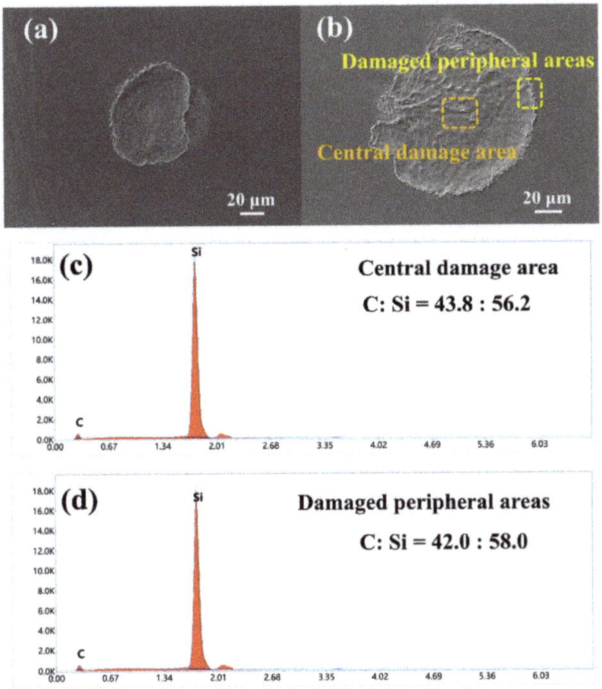

Figure 6. Damage morphologies of each SiC crystal by SEM: (**a**) HP-SiC; (**b**) N-SiC; the ratio value of different damage position of N-SiC crystal by EDS: (**c**) central damage area; (**d**) damaged peripheral area.

Table 9. The initial damage size of HP-SiC crystal and N-SiC crystal under the same incident laser fluence irradiations.

	HP-SiC	N-SiC
Damage depth (Average value)	0.87 μm	0.89 μm
The lateral damage size (Average value)	82.36 μm	124.17 μm

Subsequently, Si:C ratios at five distinct damage sites across various SiC crystals were recorded using an energy spectrometer (see Table 10). The results indicate that the Si:C ratios in the central damaged regions of both crystals are comparable to those in the undamaged areas. This phenomenon can be attributed to the Gaussian distribution of the incident laser energy used in the damage threshold testing, which concentrates higher energy at the center. Consequently, during the induction of damage in the SiC crystals, the interaction between surface carbon and atmospheric oxygen leads to the formation and subsequent loss of carbon dioxide, while the high-energy laser induces carbon agglomeration at the surface. As a result, no significant depletion of carbon occurs.

Table 10. Ratio of C and Si elements at different damage locations in HP-SiC and N-SiC.

C:Si	HP-SiC	N-SiC
Central damage area	44.7:55.3	43.4:56.6
Damaged peripheral areas	42.5:57.5	41.4:58.6

In the damaged edge regions, the ablation process has just begun. Carbon within the crystal reacts with atmospheric oxygen, leading to the significant loss of carbon via gaseous byproducts such as carbon dioxide. Analysis of the carbon-to-silicon (C:Si) element ratio in regions subjected to identical incident laser energy densities shows that the reduction in carbon content is markedly greater in the damage margins of N-SiC crystals than in those of HP-SiC crystals. Consequently, under equivalent irradiation conditions, N-SiC exhibits more severe ablation damage, with a higher proportion of its carbon combining with oxygen for removal. This reaction constitutes one of the primary factors underlying its reduced damage threshold.

5. Conclusions

This paper addresses the challenges of precision subtractive machining in hard and brittle SiC crystals. It examines the damage characteristics and mechanisms imparted by fundamental-frequency nanosecond pulsed lasers on both HP-SiC and N-SiC crystals. A testing platform for evaluating damage thresholds using fundamental-frequency nanosecond pulsed lasers was first established. Utilizing the internationally standardized 1-ON-1 laser damage threshold method, the single-pulse laser damage thresholds for both types of SiC crystals were determined. The results indicate that, when subjected to fundamental-frequency nanosecond pulsed lasers, N-SiC crystals exhibit a lower LIDT compared to the HP-SiC crystal. Subsequently, spectral absorption measurements were conducted to analyze the transmittance of two types of SiC crystals. The test results indicated that the N-SiC crystal exhibits higher absorption than HP-SiC, which partially accounts for the observed disparity in damage thresholds. In addition, theoretical calculations using a defect analysis model were performed on both crystals, revealing distinct defect characteristics that further contribute to the lower damage threshold of N-SiC crystal. A photothermal weak absorption platform was also established to characterize the subtle photothermal absorption properties of both SiC crystals, yielding corresponding absorption profiles. Finally, scanning electron microscopy and energy-dispersive spectroscopy were employed to examine damage morphologies, uncovering distinct damage characteristics and underlying physical processes. This research advances precision control techniques for efficient nanosecond pulsed laser ablation processing of various SiC crystals.

Author Contributions: Conceptualization, C.S. and T.W.; Methodology, T.W.; Software, P.X. and E.W.; Validation, P.X. and D.J.; Formal analysis, T.W.; Investigation, E.W.; Resources, D.J., Q.W. and Z.S.; Data curation, C.S., E.W., H.K., D.J. and Z.S.; Writing—original draft, C.S., P.X. and T.W.; Writing—review & editing, C.S., P.X. and E.W.; Visualization, Y.G.; Supervision, X.Z., H.K., D.J., Q.W., Z.S. and Y.G.; Project administration, X.Z., D.J., Q.W., Z.S. and Y.G.; Funding acquisition, X.Z., H.K., Q.W., Z.S. and Y.G. All authors have read and agreed to the published version of the manuscript.

Funding: This research was funded by the National Key Research and Development Program of China, grant number 2022YFB3605800.

Data Availability Statement: Data underlying the results presented in this paper are not publicly available at this time but may be obtained from the authors upon reasonable request.

Conflicts of Interest: The authors declare no conflicts of interest.

References

1. Cheng, L.; Yang, J.Y.; Zheng, W. Bandgap, mobility, dielectric constant, and Baliga's figure of merit of 4H-SiC, GaN, and β-Ga_2O_3 from 300 to 620 K. *ACS Appl. Electron. Mater.* **2022**, *4*, 4140–4145. [CrossRef]
2. Weitzel, C.E.; Moore, K.E. Performance comparison of wide bandgap semiconductor rf power devices. *J. Electron. Mater.* **1998**, *27*, 365–369. [CrossRef]
3. Park, C.H.; Cheong, B.H.; Lee, K.H.; Chang, K.J. Structural and electronic properties of cubic, 2H, 4H, and 6H SiC. *Phys. Rev. B* **1994**, *49*, 4485–4493. [CrossRef]
4. Qian, X.; Jiang, P.Q.; Yang, R.G. Anisotropic thermal conductivity of 4H and 6H silicon carbide measured using time-domain thermoreflectance. *Mater. Today Phys.* **2017**, *3*, 70–75. [CrossRef]

5. Wei, R.S.; Song, S.; Yang, K.; Cui, Y.; Peng, Y.; Chen, X.; Hu, X.; Xu, X. Thermal conductivity of 4H-SiC single crystals. *J. Appl. Phys.* **2013**, *113*, 053503. [CrossRef]
6. Itoh, A.; Matsunami, H. Single crystal growth of SiC and electronic devices. *Crit. Rev. Solid State Mater. Sci.* **1997**, *22*, 111–197. [CrossRef]
7. Roschke, M.; Schwierz, F. Electron mobility models for 4H, 6H, and 3C SiC. *IEEE Trans. Electron. Devices* **2001**, *48*, 1442–1447. [CrossRef]
8. DeVault, C.T.; Deckoff-Jones, S.; Liu, Y.; Hammock, I.N.; Sullivan, S.E.; Dibos, A.; Sorce, P.; Orcutt, J.; Awschalom, D.D.; Heremans, F.J.; et al. Silicon-on-silicon carbide platform for integrated photonics. *Adv. Opt. Mater.* **2024**, *12*, 2401101. [CrossRef]
9. Itas, Y.S.; Alsuhaibani, A.M.; Refat, M.S.; Alrahili, M.R. Envisaging the quantum capacitance in modified monolayer silicon carbide. *Appl. Phys. A* **2024**, *130*, 668. [CrossRef]
10. Chen, Y.; Liu, Z.H.; Xiao, Z.M.; Li, Z.; Hou, M.; Chen, X.; Gao, J.; Wong, C.-P. Processing, property modulation and application of one-dimensional SiC nanostructure field emitters. *Microelectron. Eng.* **2023**, *277*, 112019. [CrossRef]
11. Grinchuk, P.S.; Abuhimd, H.M.; Kiyashko, M.V.; Solovei, D.; Akulich, A.; Stepkin, M.; Lapitskaya, V.; Kuznetsova, T.; Danilova-Tretiak, S.; Evseeva, L. Advanced reaction-bonded SiC ceramics for space mirror blanks. *J. Manuf. Process.* **2024**, *113*, 275–290. [CrossRef]
12. Michaud, J.F.; Portail, M.; Alquier, D.; Certon, D.; Dufour, I. Silicon-carbide-based MEMS for gas detection applications. *Mater. Sci. Semicond. Process.* **2024**, *171*, 107986. [CrossRef]
13. Wu, R.; Chen, H.; Zhou, Y.; Guo, Y.; Ji, Z.; Li, L.; Yang, Y.; Wang, G.; Zhou, J.; Fu, Y. Advances in silicon carbides and their MEMS pressure sensors for high temperature and pressure applications. *ACS Appl. Mater. Interfaces* **2025**, *17*, 26117–26155. [CrossRef]
14. Lu, D.; Xiao, Y.; Hu, Q.; Wang, X.; Wang, Y.; Wu, R. Hydrophobic texture screening of laser ablated silicon carbide surface and its influencing factors analysis and optimization. *Adv. Eng. Mater.* **2023**, *25*, 2201733. [CrossRef]
15. Zheng, Q.Z.; Fan, Z.J.; Jiang, G.D.; Pan, A.; Yan, Z.; Lin, Q.; Cui, J.; Wang, W.; Mei, X. Mechanism and morphology control of underwater femtosecond laser microgrooving of silicon carbide ceramics. *Opt. Express* **2019**, *27*, 26264–26280. [CrossRef]
16. Zhai, Z.Y.; Wang, W.J.; Mei, X.S.; Li, M.; Cui, J.; Wang, F.; Pan, A. Effect of the surface microstructure ablated by femtosecond laser on the bonding strength of EBCs for SiC/SiC composites. *Opt. Commun.* **2018**, *424*, 137–144. [CrossRef]
17. Liu, Y.S.; Zhang, R.H.; Li, W.N.; Wang, J.; Yang, X.; Cheng, L.; Zhang, L. Effect of machining parameter on femtosecond laser drilling processing on SiC/SiC composites. *Int. J. Adv. Manuf. Technol.* **2018**, *96*, 1795–1811. [CrossRef]
18. Ransom, E.H.; Dowling, K.M.; Rocca-Bejar, D.; Palko, J.W.; Senesky, D.G. High-throughput pulsed laser manufacturing etch process for complex and released structures from bulk 4H-SiC. In Proceedings of the 2017 IEEE 30th International Conference on Micro Electro Mechanical Systems (MEMS), Las Vegas, NV, USA, 22–26 January 2017; pp. 671–674.
19. Zhang, R.; Huang, C.Z.; Wang, J.; Zhu, H.; Yao, P.; Feng, S. Micromachining of 4H-SiC using femtosecond laser. *Ceram. Int.* **2018**, *44*, 17775–17783. [CrossRef]
20. Wu, C.; Fang, X.D.; Kang, Q.; Sun, H.; Zhao, L.; Tian, B.; Fang, Z.; Pan, M.; Maeda, R.; Jiang, Z. Crystal cleavage, periodic nanostructure and surface modification of SiC ablated by femtosecond laser in different media. *Surf. Coat. Technol.* **2021**, *424*, 127652. [CrossRef]
21. Chen, J.B.; Chen, X.X.; Zhang, X.H.; Zhang, W. Effect of laser incidence angle on the femtosecond laser ablation characteristics of silicon carbide ceramics. *Opt. Laser Eng.* **2024**, *172*, 107849. [CrossRef]
22. Tomita, T.; Kinoshita, K.; Matsuo, S.; Hashimoto, S. Effect of surface roughening on femtosecond laser-induced ripple structures. *Appl. Phys. Lett.* **2007**, *90*, 153115. [CrossRef]
23. Shen, J.; Chen, H.; Chen, J.; Lin, L.; Gu, Y.; Jiang, Z.; Li, J.; Sun, T. Mechanistic difference between Si-face and C-face polishing of 4H–SiC substrates in aqueous and non-aqueous slurries. *Ceram. Int.* **2023**, *49*, 7274–7283. [CrossRef]
24. *ISO 21254–1*; Lasers and Laser-Related Equipment—Test Methods for Laser-Induced Damage Threshold. ISO: Geneva, Switzerland, 2011.
25. Krol, H.; Gallais, L.; Grezes-Besset, C.; Natoli, J.-Y.; Commandré, M. Investigation of nanoprecursors threshold distribution in laser-damage testing. *Opt. Commun.* **2005**, *256*, 184–189. [CrossRef]
26. Natoli, J.Y.; Gallais, L.; Akhouayri, H.; Amra, C. Laser-induced damage of materials in bulk, thin-film, and liquid forms. *Appl. Opt.* **2002**, *41*, 3156–3166. [CrossRef] [PubMed]

photonics

MDPI

Article

Near-Infrared Absorption Enhancement of GaAs Photocathode Through "Sandwich" Micro-Nano Structure

Ziyang Xiao [1,2], Miao Dong [1,3], Yonggang Huang [1,3,4,*], Jinhui Yang [1,3], Peng Jiao [1,3], Pan Shi [1,5], Yajie Du [1,3], Ying He [6], Jing Cheng [1,2] and Yinsheng Xu [2,*]

1 China Building Materials Academy, Beijing 100024, China
2 State Key Laboratory of Silicate Materials for Architectures, Wuhan University of Technology, Wuhan 430070, China
3 Key Laboratory of Special Optoelectronic Glass in Building Materials Industry, China Building Materials Academy, Beijing 100024, China
4 Glass-Based Functional Material Technology Innovation Center, China Building Materials Academy, Beijing 100024, China
5 China National Building Material Guangxin Technology Co., Ltd., Zaozhuang 277101, China
6 College of General Education, Xi'an Mingde Institute of Technology, Xi'an 710124, China
* Correspondence: huangyonggang@cbma.com.cn (Y.H.); xuyinsheng@whut.edu.cn (Y.X.)

Abstract

In this paper, a nano-layered transmission GaAs photocathode structure is proposed. The near-infrared absorption of the photocathode is enhanced by inserting a "sandwich" structure of nano-SiO_2 layer + Si_3N_4 nanopillar array + nano-SiO_2 layer between the cathode optical window and the photocathode. Compared with the flat film structure GaAs photocathode used in the current third-generations image intensifiers, the optical absorption of the optimized "sandwich" structure GaAs photocathode in the near-infrared band has been significantly improved: when the wavelength λ is 868 nm and 896 nm, the optical absorption is increased by 41.69%, 55.08%, respectively. The effects of structural parameters including film thickness and grating filling medium on the light absorption of photocathode are investigated. The results show that the near-infrared light absorption enhancement is the most obvious when Si_3N_4 is selected as the grating filling medium for the current design, and the deposition of SiO_2 film with 10 nm thickness could effectively prevent the damage of Si_3N_4 during bonding with the photocathode. The theoretical analyses offer important guidance in material selection and structural optimization in the grating cathode optical window used in the third-generation image intensifier for improving performance.

Keywords: nanopillar array; GaAs photocathode; guided-mode resonance; optical performance

1. Introduction

Gallium arsenide negative electron affinity (NEA) photocathodes are the core photoelectric conversion components of third-generation image intensifiers. They enable efficient photoelectron excitation and emission, featuring high quantum efficiency (QE), low dark emission, narrow energy and angular distributions of emitted electrons, long-wavelength tunability, and great potential for wavelength response extension. Owing to these advantages, they are widely applied in spin-polarized electron sources, photodetectors, and low-light-level imaging systems [1–5]. In application scenarios of low-light-level night vision devices, photocathodes typically detect the reflected light of a scene. Taking grassland as a typical example, the reflected spectral energy of night-sky illumination is generally weak in the visible band but concentrated in the near-infrared region [6]. For conventional planar GaAs photocathodes, the intrinsic absorption coefficient and thickness limitations cause a rapid decrease in absorption efficiency in the long-wavelength region, resulting in significantly reduced quantum efficiency. This severely restricts the near-infrared detection capability of photoelectric imaging systems under extremely low illumination.

According to Spicer's three-step model of photoemission [7], increasing the absorption of incident photons, enhancing photoelectron transport, and reducing the electron emission work function are three main approaches to improving photocathode quantum efficiency. Graded-doping or graded-bandgap structures [8–11] bend the energy bands and form internal built-in electric fields. These built-in fields facilitate electron transport toward the emitting surface, increasing their diffusion length and enhancing the spectral response of the photocathode. Chen et al. [2] grew a transmission-mode (t-mode) AlGaAs photocathode with a graded Al-composition structure using molecular beam epitaxy (MBE). When the outermost GaAs layer reached nano-scale, the quantum efficiency in the blue-green spectral region increased significantly. This is because nano-scale surface structures exhibit quantum confinement effects, which reduce the surface electron affinity and enhance the electron escape probability, thereby allowing more photoelectrons to be emitted into the vacuum. Enhancing optical absorption is also a common method for improving quantum efficiency. Typical structures include distributed Bragg reflector (DBR) structures and nano-array structures. Wang et al. [12] utilized a DBR to reflect specific wavelengths of transmitted light back into the emission layer to further enhance absorption, thereby improving the response of AlGaAs photocathodes at 532 nm. For t-mode multi-alkali photocathodes used in super-second-generation image intensifiers, a transmissive diffraction grating on the glass window to increase the optical path within the emission layer was proposed [13]. It is reported that the visible light and near-infrared light absorption is effectively enhanced without altering the photocathode thickness. With regard to negative electron affinity (NEA) GaAs photocathodes, existing studies indicate that, for both reflection-mode (r-mode) [14–16] and transmission-mode photocathodes [17], the prevailing approach is to nanostructure one of the photocathode layers. By contrast, design strategies that do not modify the intrinsic photocathode structure, similar to those adopted in super-second-generation image intensifiers, have only been reported for nano-scale NEA photocathodes [18].

In this work, a "sandwich" t-mode NEA GaAs photocathode structure consisting of a nano-SiO_2 layer, a Si_3N_4 nanopillar array (NPA), and another nano-SiO_2 layer inserted between the photocathode window was proposed. This configuration aims to enhance near-infrared (800–900 nm) absorption in GaAs photocathode. An optical model based on the finite-difference time-domain (FDTD) method was established, in which the photocathode structure is the same as the material composition and architecture of existing third-generation t-mode NEA GaAs photocathodes. The relationship between structural parameters of the nanopillar array and the 800–900 nm absorption of the photocathode was investigated. By adjusting these parameters, significant absorption enhancement was achieved. Additionally, the effects of film thickness and grating filling material on optical absorption were analyzed. These theoretical analyses that could be verified by the other applicable fields serve as the most important guidance for material selection and structural design of grating-type cathode windows for third-generation image intensifiers.

2. Materials and Methods

The unit cell of the designed "sandwich" nanostructured GaAs photocathode is shown in Figure 1a. The device is mainly comprised of a grating-type cathode glass window and a t-mode GaAs photocathode. The grating-type cathode glass window consists of an anti-

 https://doi.org/10.3390/photonics13010079

halo photoelectric glass substrate, a SiO_2 layer, a hexagonal close-packed Si_3N_4 nanopillar array with its fill layer, and thin SiO_2 nano-layers. The GaAs photocathode stack includes a SiO_2/Si_3N_4 passivation layer, an $Al_{0.7}Ga_{0.3}As$ window layer, an $Al_{0.2}Ga_{0.8}As$ window layer, and the GaAs emitting layer. The nanopillar array is characterized with period P, pillar diameter D, and pillar height H. Compared with approaches that directly nanostructure the photocathode itself, it is unnecessary for the present design to consider the negative feedback of electron collection efficiency on quantum efficiency [15,19]. Without altering the structure of the GaAs photocathode, the photocathode optical absorption can be controlled by adjusting the structural parameters of the nanopillar array.

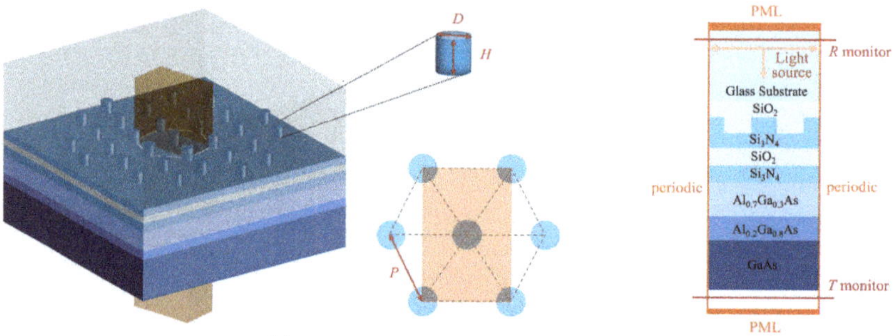

Figure 1. Construction of structural model for "sandwich" nanostructured GaAs photocathode. (**a**) Structural unit of GaAs photocathode inserted into "sandwich" nanostructured layer; (**b**) FDTD simulation region and monitor settings.

Optical modeling and simulations of the GaAs photocathode were conducted based on the finite-difference time-domain (FDTD) method. The simulation region and monitor arrangement are illustrated in Figure 1b. The horizontal boundary conditions of the simulation region were set to periodic, and the vertical boundary conditions were set to a perfectly matched layer (PML). The light source was a normally incident plane wave with a wavelength range of 500–1000 nm. The reflection (R) and transmission (T) were monitored by placing two-dimensional power monitors on the x–y plane in the SiO_2 window and vacuum, respectively. The reflection monitor (R monitor) was placed above the light source, and the transmission monitor (T monitor) was placed below the emission layer. The absorption (A) was calculated by $A = 1 - R - T$. The refractive indices and extinction coefficients for materials such as Si_3N_4 and GaAs were taken from references [20–27]. For the Si_3N_4 material, the main fitting parameters include fit tolerance 0.03 and max coefficient 6. The simulation mesh was set with mesh accuracy 3, and the nanopillar array region was subjected to mesh refinement, with a step size of 5 nm in the x, y, and z directions; i.e., $dx = dy = dz = 5$ nm.

3. Results and Discussion

3.1. Optimization of Nanopillar Array Structure

Focusing on achieving high absorption in the near-infrared short-wave range for the GaAs photocathode, the structure of the nanopillar array was firstly optimized. A parameter sweep of the Si_3N_4 nanopillar array structure was performed with a scanning interval of 10 nm, and the optimal structural parameters were determined as $P = 680$ nm, $D = 400$ nm, and $H = 310$ nm. The optical performance of the photocathode with this optimized structure is shown in Figure 2. In comparison with the typical flat film GaAs photocathodes used in third-generation image intensifiers, the nanopillar array photocath-

 https://doi.org/10.3390/photonics13010079

ode exhibits a significant enhancement in absorption within the near-infrared band. In particular, at wavelengths of 868 nm, 896 nm, and 957 nm, the absorption is increased by 41.69%, 55.08%, and 18.86%, respectively. Correspondingly, Figure 2b shows that the reflection at these wavelengths decreases substantially, while the transmission increases to varying degrees. Since the wavelength λ = 957 nm exceeds the cutoff wavelength of GaAs photocathodes, the enhancement at this wavelength is of limited practical significance and not be discussed further. To investigate the mechanism by which this nanostructure enhances light absorption, the magnetic field intensity distributions at λ = 868 nm and 896 nm were analyzed through the vertical cross-section of the Si_3N_4 nanopillars, as shown in Figure 3. Local magnetic field hot spots are observed to be periodically distributed along the x-axis within the Si_3N_4 nanopillars. This distribution matches the characteristic features of guided-mode resonances excited in waveguide gratings [28], indicating that guided-mode resonance is excited in the nanopillar array structure.

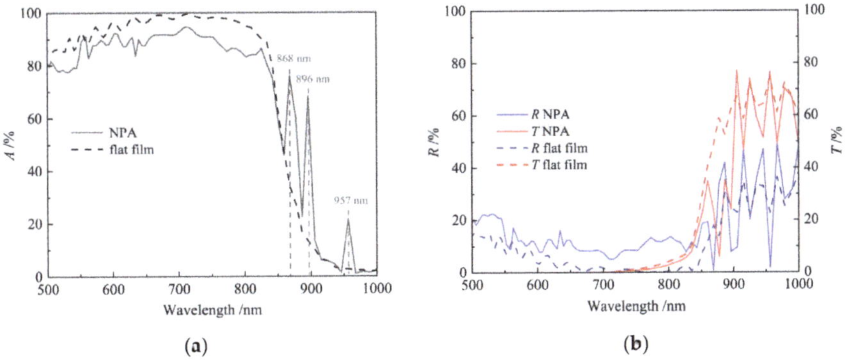

Figure 2. Optical properties of Si_3N_4 nanopillar array (NPA) and flat film structured GaAs photocathode. (**a**) Absorption spectra; (**b**) reflection and transmission spectra.

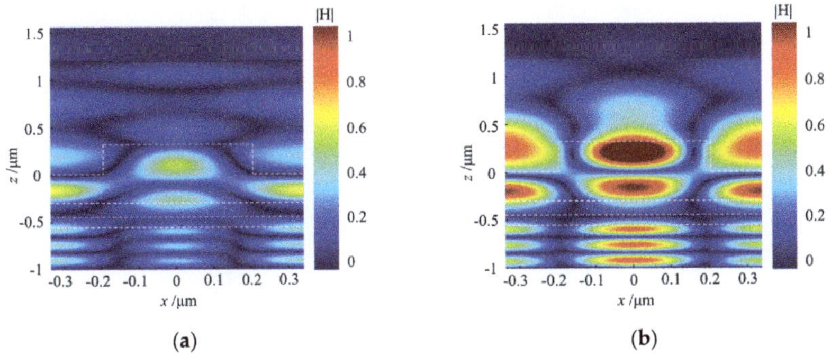

Figure 3. Magnetic field intensity distribution through the vertical cross−section of the Si_3N_4 nanopillar, in which the dashed line is the outline of nanopillar and each film layer. (**a**) λ = 868 nm; (**b**) λ = 896 nm.

When incident light interacts with the grating, multiple diffraction orders are generated. Guided-mode resonance occurs when the wave vector of a diffracted order matches the phase of the guided-mode wave vector, causing lateral propagation along the grating surface (the x-y plane) and confining the energy within the waveguide layer [29,30]. During propagation, the guided-mode experiences leakage due to the periodic perturbation of the grating. The wave leaking upwards undergoes destructive interference with the reflected wave, and the wave leaking downwards undergoes constructive interference with the

https://doi.org/10.3390/photonics13010079

transmitted wave. This results in a decrease in reflection and an increase in transmission. Consequently, the optical energy is efficiently directed toward the transmission side and absorbed by the underlying photocathode layers, thereby enhancing the absorption of the photocathode. In the subwavelength regime, only the zeroth-order diffracted light can propagate to the far field, while higher-order diffracted waves become evanescent. Thus, energy is exchanged solely between the zeroth-order reflection and transmission channels, leading to sharp reflection or transmission peaks at the resonance wavelengths. To further investigate this effect, the extinction coefficient of the absorbing material in the model was adjusted to avoid absorption, enabling observation of the intrinsic reflection and transmission characteristics of the nanostructure. At λ = 868 nm and 896 nm, the refractive index of GaAs is 3.54, and only the extinction coefficient k of GaAs is nonzero. For GaAs, the refractive index n was set to 3.54 and the extinction coefficient k was set to 0. The simulated optical response under these conditions is shown in Figure 4. Transmission and reflection peaks appear at 860 nm and 905 nm, respectively, further confirming the presence of guided-mode resonance. The slight deviation from 868 nm and 896 nm may be attributed to approximations in the material refractive index and extinction coefficient.

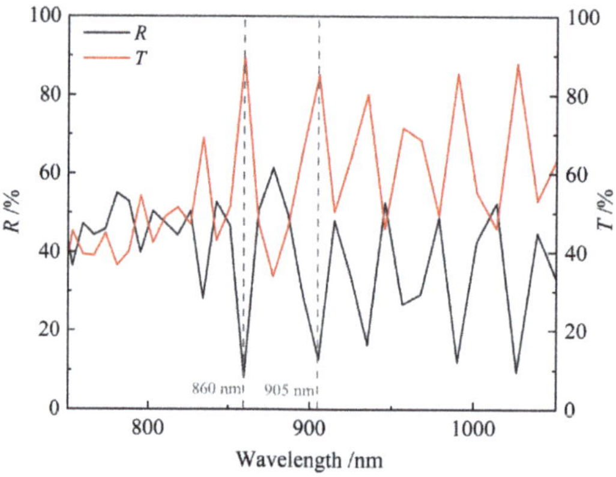

Figure 4. Reflection and transmission spectra of nanopillar array photocathode at n_{GaAs} = 3.54 and k_{GaAs} = 0.

3.2. *Influence of Varied Nanopillar Array Structure on Photocathode Absorption*

During the fabrication of nanostructures, the structural parameters generally show a gap from the theoretical design due to inevitable errors introduced by lithography and etching. To determine the acceptable tolerance range, the effects of each structural parameter of the nanopillar array on the photocathode absorption were investigated based on the modelling optimized structure. Figure 5 shows the absorption spectra of photocathodes with different array periods. As the nanopillar period increases, the number, position, and peak values of the absorption peaks all change. Specifically, at λ = 868 nm and 896 nm, the peak absorption first increases and then decreases as the period increases, with a large amplitude of change. Existing studies on the influence of waveguide structural parameters on guided-mode resonance mainly focus on single-layer or double-layer waveguide grating structures. However, the nanostructured photocathode model in this work is a multilayer waveguide grating structure. The influence of the nanostructure parameters on photocathode absorption differs from the trends predicted by the conventional guided-mode resonance theory. This difference could be attributed to the mutual influence between guided-mode resonance and effects such as thin-film interference. The effects of nanopillar diameter and height on the photocathode absorption are shown in Figure 6, and they exhibit similar trends. Taking diameter as an example, the peak absorption at λ = 868 nm and 896 nm

https://doi.org/10.3390/photonics13010079

first increases and then decreases as the diameter increases. At λ = 868 nm, the peak varies slowly with diameter, resulting in a wider bandwidth for the peak absorption–diameter curve. At λ = 896 nm, the variation is faster and the curve is narrower.

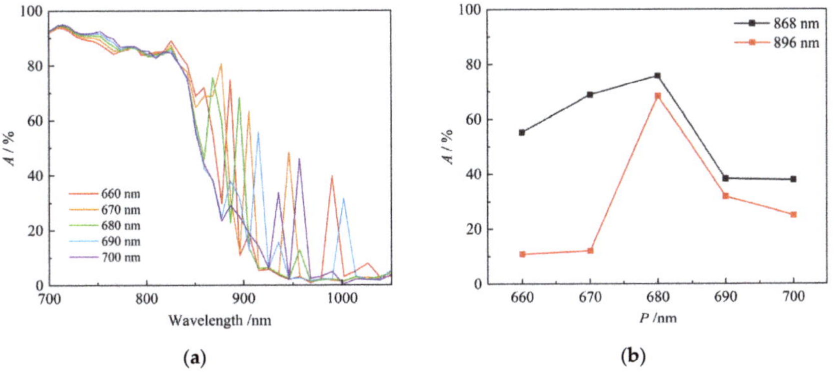

Figure 5. The relationship between the period of nanopillar array and the light absorption of the photocathode. (**a**) Photocathode absorption spectra at different periods; (**b**) variation in peak absorption with period at λ = 868 nm and 896 nm.

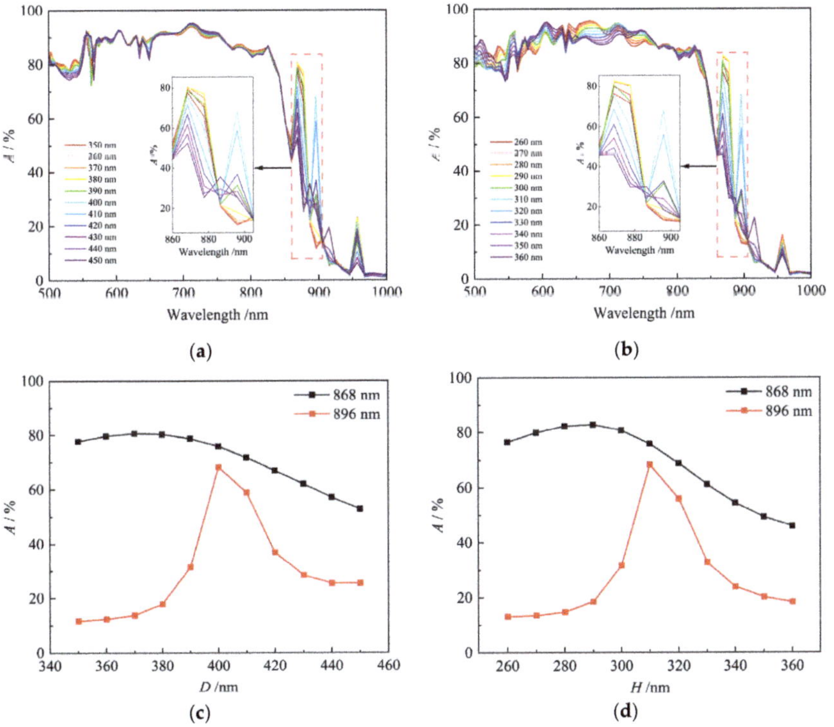

Figure 6. The relationship between the structural parameters and the light absorption of the photocathode. (**a**) Photocathode absorption spectra at different nanopillar diameters; (**b**) photocathode absorption spectra at different nanopillar heights; (**c**,**d**) variation in peak absorption with diameter and height at λ = 868 nm and λ = 896 nm.

https://doi.org/10.3390/photonics13010079

3.3. *Influence of Refractive Index of the Grating Filling Material on Photocathode Absorption*

Keeping the structural parameters unchanged, the effect of the filling material's refractive index on the photocathode absorption was investigated. The filling material Si_3N_4 ($n \sim 2.06$) was modified to other dielectrics including Al_2O_3 ($n \sim 1.68$), HfO_2 ($n \sim 1.89$), Ta_2O_5 ($n \sim 2.13$), and TiO_2 ($n \sim 2.5$), and the comparative results are shown in Figure 7. The refractive index has a significant regulatory effect on the photocathode's light absorption, with clear performance differences among the various grating filling materials. In the visible light band and parts of the near-infrared short-wave region ($\lambda < 850$ nm), the absorption of the grating photocathodes with all filling materials is lower than that of the flat film photocathode. When λ is over 850 nm, however, the grating structure photocathodes in all groups exhibit several absorption peaks, showing a distinct enhancement compared to the flat film photocathode. Furthermore, the peak positions and values of the absorption peaks vary among the different filling materials. According to the optical absorptivity in the near-infrared short-wave region and considering the cut-off wavelength of the GaAs photocathode, the integral absorptivity of the photocathode in the 850–950 nm band under different filling materials is calculated (Table 1), and it is concluded that Si_3N_4 is the best grating filling material for optical absorption enhancement.

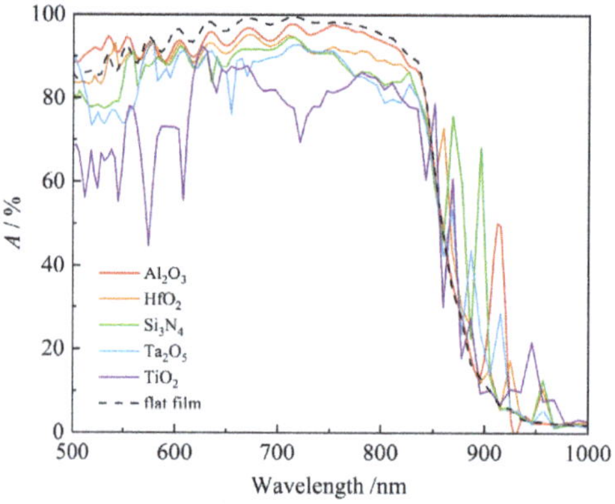

Figure 7. Influence of grating filling materials on light absorption of photocathode.

Table 1. Integrated absorptivity of photocathodes in 850–950 nm band with different grating filling materials.

Material	Integrated Absorptivity in 850–950 nm Band
Al_2O_3	25.35%
HfO_2	24.09%
Si_3N_4	31.18%
Ta_2O_5	25.68%
TiO_2	22.97%
Flat film	18.54%

3.4. *Influence of SiO_2 Thickness on Photocathode Absorption*

Considering the requirements of practical fabrication processes, in order to prevent the Si_3N_4 from being damaged during the bonding of the photocathode optical window with the GaAs epitaxial wafer, a SiO_2 thin film is deposited on the surface of Si_3N_4 filling material. This allows the bonding process to occur between two SiO_2 surfaces, improving material compatibility and process adaptability. Therefore, the influence of the SiO_2 film thickness (d_1) on the optical absorption was investigated, and the results are shown in Figure 8. As the film thickness increases, the absorption peak at $\lambda = 868$ nm decreases, while

 https://doi.org/10.3390/photonics13010079

the peak at λ = 896 nm first increases and then decreases. Based on the simulation results, the integrated absorption in the near-infrared region indicates that the system achieves the optimum absorption and protection from damage when the SiO_2 film thickness is 180 nm. Since the GaAs epitaxial wafer used in third-generation image intensifiers already contains a native 170 nm thick SiO_2 layer on its surface, a 10 nm thick SiO_2 film can be deposited on the Si_3N_4 during grating fabrication. This configuration not only achieves the optimal optical absorption but also protects the Si_3N_4 filling material from structural damage during the bonding process.

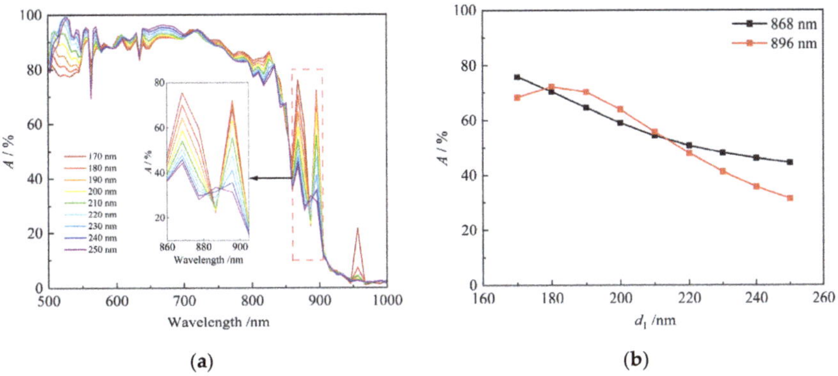

Figure 8. The relationship between the thickness of SiO_2 layer and the absorption spectrum of photocathode. (**a**) The absorption spectra of photocathode with different d_1; (**b**) variation in peak absorption with d_1 at λ = 868 nm and λ = 896 nm.

3.5. Discussion

This design focuses on the GaAs photocathode used in third-generation image intensifiers. Without altering the original photocathode structure, enhanced optical absorption in specific wavelength bands is achieved. Meanwhile, micro-nano fabrication processes are taken into consideration, and corresponding investigations and optimizations are conducted to facilitate practical fabrication and production. At present, the commonly adopted approach for nanostructured NEA GaAs photocathodes is to directly pattern the photocathode surface. Taking a Mie-type nanopillar array reflective GaAs photocathode [15] as an example, the generated photoelectrons are highly localized inside the nanopillars, mainly near the top surfaces and sidewalls. However, not all of these photoelectrons can be effectively transported and emitted into the vacuum. In the absence of space-charge effects and external electromagnetic fields, only electrons emitted within a limited angular range can be collected and thus contribute positively to the quantum efficiency. Electrons emitted outside this angular range are likely to be trapped by the nanopillar sidewalls (e.g., laterally emitted electrons), which leads to a negative impact on the QE. In contrast, the proposed design preserves the planar structure of the GaAs photocathode. As a result, issues related to electron collection efficiency and its negative feedback on the quantum efficiency do not need to be considered.

4. Conclusions

This study proposes a nanostructured t-mode GaAs NEA photocathode in which a Si_3N_4 nanopillar array grating is employed to enhance optical absorption while preserving the original photocathode structure and thereby improve photoemission performance. By reproducing the composition and configuration of the t-mode GaAs NEA photocathodes used in existing third-generation image intensifiers, the optimized structural parameters of nanopillar array as P = 680 nm, D = 400 nm, and H = 310 nm for the modelled photocathode

https://doi.org/10.3390/photonics13010079

structure are achieved. The significant absorption enhancement in the near-infrared region, especially at characteristic wavelengths of 868 nm and 896 nm, are realized, showing the absorption increases by 41.69% and 55.08% in comparison with a typical flat film GaAs photocathode. Fully considering the micro-nano fabrication processes in practice, the structural parameters including the period, diameter, and height of the nanopillar array are adjusted to investigate the influence on optical absorption. It is shown that the period should be focused on more in controlling the optimum value. According to the comparison with the other grating filling materials, Si_3N_4 is indeed the most effective filling material for enhancing near-infrared absorption with the designed structure parameters due to the suitable refractive index. During grating fabrication with Si_3N_4 as the filling material, a 10 nm SiO_2 layer may be deposited on the Si_3N_4 surface to protect the Si_3N_4 from structural damage during the bonding process without negatively impacting the light absorption performance. In future work, grating-type cathode optical window samples will be fabricated according to the optimized parameters and bonded to GaAs photocathodes.

Author Contributions: Conceptualization, J.Y. and Y.H. (Yonggang Huang); methodology, Z.X.; formal analysis, M.D.; writing—original draft, Z.X.; writing—review and editing, Y.H. (Yonggang Huang) and Y.X.; supervision, P.S., Y.H. (Ying He) and J.C.; project administration, P.J. and Y.D.; funding acquisition, Y.H. (Yonggang Huang). All authors have read and agreed to the published version of the manuscript.

Funding: This work is supported by the National Natural Science Foundation of China (No. 52072357), the Young Elite Scientists Sponsorship Program by CAST, the Youth Talent Support Project of Beijing Association for Science and Technology (No.BYESS2024388), the Major Scientific and Technological Research Projects in the National Building Materials Industry of the China Building Materials Federation (No.202201JBGS11-01), and the Key Core Technology Research Projects of China National Building Materials Group (No.2023SJYL07).

Institutional Review Board Statement: Not applicable.

Informed Consent Statement: Not applicable.

Data Availability Statement: Data underlying the results presented in this paper are not publicly available at this time but may be obtained from the authors upon reasonable request.

Conflicts of Interest: Author Ziyang Xiao, Miao Dong, Yonggang Huang, Jinhui Yang, Peng Jiao, Pan Shi, Yajie Du, Jing Cheng were employed by the company China Building Materials Academy. Author Pan Shi was also were employed by the company China National Building Material Guangxin Technology Co., Ltd. The remaining authors declare that the research was conducted in the absence of any commercial or financial relationships that could be construed as a potential conflict of interest.

References

1. Zhao, J.; Yu, H.; Liu, W.; Guo, J. Analysis of the Relation between Spectral Response and Absorptivity of GaAs Photocathode. *Acta Phys. Sin.* **2017**, *66*, 227801. [CrossRef]
2. Chen, X.; Tang, G.; Wang, D.; Xu, P. High Quantum Efficiency Transmission-Mode GaAlAs Photocathode with a Nanoscale Surface Structure. *Opt. Mater. Express* **2018**, *8*, 3155. [CrossRef]
3. Zhang, Y. Progress in Research on Semiconductor Photocathodes. *Infrared Technol.* **2022**, *44*, 778–791.
4. Cai, Z.; Huang, W.; Yang, C.; Lou, B.; He, J. Theoretical Study of the Average Decay Time and Response Characteristics of the Transmission-Mode GaAs Photocathodes with the Different Doping Gradient Distribution. *Chin. J. Vac. Sci. Technol.* **2025**, *45*, 398–407. [CrossRef]
5. Xie, H. Overview of the Semiconductor Photocathode Research in China. *Micromachines* **2021**, *12*, 1376. [CrossRef]
6. Zhang, Q.; Bai, X.; Cheng, H.; Jiao, G.; Li, Z.; Han, K.; Li, Q. Research on Signal-to-noise Ratio of Low-light Level Image Intensifier Based on Night Sky Light Spectrum Matching. *Acta Phys. Sin.* **2022**, *51*, 0304005. [CrossRef]
7. Spicer, W.E. Photoemissive, Photoconductive, and Optical Absorption Studies of Alkali-Antimony Compounds. *Phys. Rev.* **1958**, *112*, 114–122. [CrossRef]

https://doi.org/10.3390/photonics13010079

8. Zhang, Y.; Chang, B.; Niu, J.; Zhao, J.; Zou, J.; Shi, F.; Cheng, H. High-Efficiency Graded Band-Gap $Al_xGa_{1-x}As/GaAs$ Photocathodes Grown by Metalorganic Chemical Vapor Deposition. *Appl. Phys. Lett.* **2011**, *99*, 101104. [CrossRef]
9. Zhang, Y.; Niu, J.; Zhao, J.; Zou, J.; Chang, B.; Shi, F.; Cheng, H. Influence of Exponential-Doping Structure on Photoemission Capability of Transmission-Mode GaAs Photocathodes. *J. Appl. Phys.* **2010**, *108*, 093108. [CrossRef]
10. Feng, C.; Zhang, Y.; Liu, J.; Qian, Y.; Liu, X.; Shi, F.; Cheng, H. Effect of Graded Bandgap Structure on Photoelectric Performance of Transmission-Mode $Al_xGa_{1-x}As/GaAs$ Photocathode Modules. *Appl. Opt.* **2017**, *56*, 9044–9049. [CrossRef]
11. Feng, C.; Zhang, Y.; Qian, Y.; Xu, Y.; Liu, X.; Jiao, G. Quantum Efficiency of Transmission-Mode $Al_xGa_{1-x}As/GaAs$ Photocathodes with Graded-Composition and Exponential-Doping Structure. *Opt. Commun.* **2016**, *369*, 50–55. [CrossRef]
12. Wang, D.; Zhang, Y.; Li, S.; Tong, Z.; Tang, S.; Shi, F.; Jiao, G.; Cheng, H.; Fu, R.; Qian, Y.; et al. AlGaAs Photocathode with Enhanced Response at 532 nm. *Acta Phys. Sin.* **2024**, *73*, 118503. [CrossRef]
13. Nützel, G.; Lavoute, P. Semi-Transparent Photocathode with Improved Absorption Rate. U.S. Patent US20150279606A1, 1 May 2018.
14. Peng, X.; Wang, Z.; Liu, Y.; Manos, D.M.; Poelker, M.; Stutzman, M.; Tang, B.; Zhang, S.; Zou, J. Optical-Resonance-Enhanced Photoemission from Nanostructured GaAs Photocathodes. *Phys. Rev. Appl.* **2019**, *12*, 064002. [CrossRef]
15. Peng, X.; Poelker, M.; Stutzman, M.; Tang, B.; Zhang, S.; Zou, J. Mie-Type GaAs Nanopillar Array Resonators for Negative Electron Affinity Photocathodes. *Opt. Express* **2020**, *28*, 860. [CrossRef]
16. Li, S.; Zhang, Y.; Wang, Z.; Wang, D.; Tang, S.; Zhang, J.; Shi, F.; Jiao, G.; Cheng, H.; Hao, G. Enhanced Blue-Green Response of Nanoarray AlGaAs Photocathodes for Underwater Low-Light Detection. *Opt. Express* **2023**, *31*, 26014–26026. [CrossRef] [PubMed]
17. Li, S.; Zhang, Y.; Tang, S.; Yang, L.; Ma, X.; Gao, J.; Tong, Z.; Qian, Y.; Shi, F.; Guo, X. Near-Infrared Enhancement in Thin Transmission-Mode GaAs Photocathodes via Nanostructured Window Layer. *Opt. Express* **2025**, *33*, 38543–38552. [CrossRef]
18. Peng, X.; Zhong, C.; Zou, J.; Deng, W. All-Dielectric Meta-Surface Transmission-Mode Ultra-Thin GaAs Negative Electron Affinity Photocathode. *Appl. Phys. Lett.* **2024**, *125*, 071103. [CrossRef]
19. Dong, M.; Wang, L.; Chen, D.; Kang, G.; Gong, Y.; Wang, Y.; Zheng, D.; Zhao, Y.; Cai, Y. Absorption-Enhanced Nanopillar-Arrayed Na_2KSb Photocathode for Improving Image Intensifier Performance. *Opt. Express* **2025**, *33*, 36025–36036. [CrossRef]
20. Malitson, I.H. Interspecimen Comparison of the Refractive Index of Fused Silica. *J. Opt. Soc. Am.* **1965**, *55*, 1205–1208. [CrossRef]
21. Luke, K.; Okawachi, Y.; Lamont, M.R.E.; Gaeta, A.L.; Lipson, M. Broadband Mid-Infrared Frequency Comb Generation in a Si_3N_4 Microresonator. *Opt. Lett.* **2015**, *40*, 4823–4826. [CrossRef]
22. Aspnes, D.E.; Kelso, S.M.; Logan, R.A.; Bhat, R. Optical Properties of $Al_xGa_{1-x}As$. *J. Appl. Phys.* **1986**, *60*, 754–767. [CrossRef]
23. Papatryfonos, K.; Angelova, T.; Brimont, A.; Reid, B.; Guldin, S.; Smith, P.R.; Tang, M.; Li, K.; Seeds, A.J.; Liu, H.; et al. Refractive Indices of MBE-Grown $Al_xGa_{(1-x)}As$ Ternary Alloys in the Transparent Wavelength Region. *AIP Adv.* **2021**, *11*, 025327. [CrossRef]
24. Bright, T.J.; Watjen, J.I.; Zhang, Z.M.; Muratore, C.; Voevodin, A.A.; Koukis, D.I.; Tanner, D.B.; Arenas, D.J. Infrared Optical Properties of Amorphous and Nanocrystalline Ta_2O_5 Thin Films. *J. Appl. Phys.* **2013**, *114*, 083515. [CrossRef]
25. Boidin, R.; Halenkovič, T.; Nazabal, V.; Beneš, L.; Němec, P. Pulsed Laser Deposited Alumina Thin Films. *Ceram. Int.* **2016**, *42*, 1177–1182. [CrossRef]
26. Al-Kuhaili, M.F. Optical Properties of Hafnium Oxide Thin Films and Their Application in Energy-Efficient Windows. *Opt. Mater.* **2004**, *27*, 383–387. [CrossRef]
27. Siefke, T.; Kroker, S.; Pfeiffer, K.; Puffky, O.; Dietrich, K.; Franta, D.; Ohlídal, I.; Szeghalmi, A.; Kley, E.; Tünnermann, A. Materials Pushing the Application Limits of Wire Grid Polarizers Further into the Deep Ultraviolet Spectral Range. *Adv. Opt. Mater.* **2016**, *4*, 1780–1786. [CrossRef]
28. Magnusson, R.; Ko, Y.H. Guided-Mode Resonance Device Technology: Design, Fabrication, Applications and Prospects. In Proceedings of the 2017 Photonics North (PN), Ottawa, ON, Canada, 6–8 June 2017; p. 1. [CrossRef]
29. Khaleque, T.; Magnusson, R. Light Management through Guided-Mode Resonances in Thin-Film Silicon Solar Cells. *J. NanoPhotonics* **2014**, *8*, 083995. [CrossRef]
30. Fallah, H.; Hakim, L.; Boonruang, S.; Mohammed, W.S.; Hsu, S.H. Polymer-Based Guided-Mode Resonance Sensors: From Optical Theories to Sensing Applications. *ACS Appl. Polym. Mater.* **2023**, *5*, 9700–9713. [CrossRef]

 https://doi.org/10.3390/photonics13010079

photonics

MDPI

Communication

Multiband Infrared Photodetection Based on Colloidal Quantum Dot

Yingying Xu [1,2], Xiaomeng Xue [1], Lixiong Wu [3], Zhikai Gan [4,*], Menglu Chen [1,*] and Qun Hao [1]

[1] School of Optics and Photonics, Beijing Institute of Technology, Beijing 100081, China; 3420205014@bit.edu.cn (Y.X.); 3120235142@bit.edu.cn (X.X.); qhao@bit.edu.cn (Q.H.)
[2] National Institute of Metrology, Beijing 100029, China
[3] State Key Laboratory of Laser Interaction with Matter, Xi'an 710024, China; wulixiong@nint.ac.cn
[4] National Key Laboratory of Infrared Detection Technologies, Shanghai 200083, China
* Correspondence: ganzhikai@mail.sitp.ac.cn (Z.G.); menglu@bit.edu.cn (M.C.)

Abstract

Multispectral infrared detection plays a crucial role in advanced applications spanning environmental monitoring, military surveillance, and biomedical diagnostics, offering superior target identification accuracy compared to single-band imaging techniques. In this work, we synthesized four distinct bands of colloidal quantum dots (CQDs)—specifically, a cut-off of 1.3 μm with PbS CQDs and 1.8 μm, 2.6 μm, and 3.5 μm with HgTe CQDs—and employed them to construct planar multiband infrared photodetectors. The device exhibited a clear photoresponse at room temperature from 0.8 μm to 3.5 μm, with responsivity of 5.39 A/W and specific detectivity of 2.01×10^{11} Jones at 1.8 μm. This materials–device co-design strategy integrates wavelength-selective CQD synthesis with planar pixel-level patterning, providing a versatile pathway for developing low-cost, solution-processed, multiband infrared photodetectors.

Keywords: colloidal quantum dot; infrared; photodetector

1. Introduction

Infrared detection [1] identifies and analyzes targets by capturing their emitted infrared radiation. Compared to single-band detection, multi-spectral infrared detection [2] overcomes the limitations of limited information dimensions. This approach enables the acquisition of spectral data within specific wavelength ranges, providing a more complete set of radiation and reflection characteristics. Consequently, it provides deeper insights into material composition, physical states, and surface properties, with broad applications in environmental monitoring [3], military reconnaissance [4,5] and Biomedical contexts [6]. Currently, traditional multispectral detection is typically based on epitaxial materials like mercury cadmium telluride (HgCdTe) [7] and indium gallium arsenide (InGaAs) [8], which offer high sensitivity and low noise. However, their complex material growth, limited bandgap tunability, and poor compatibility with silicon readout circuits pose significant challenges for device integration and optimization.

Colloidal quantum dots (CQDs) [9,10] are promising materials for developing high-performance optoelectronic devices with a wide spectral range due to their unique properties, such as strong light absorption, multi-exciton generation, and size-tunable bandgaps. Among these colloidal nanomaterials, metal chalcogenide quantum dots, such as PbS CQDs and HgTe CQDs, have been shown to exhibit excellent photoelectric responses and broad spectral ranges. By precisely controlling the size of the CQDs, their optical response range can be flexibly tuned, providing diversified options for multi-spectral infrared detection. However, the intrinsic bandgap and quantum confinement modulation ranges of PbS and HgTe CQDs differ significantly. PbS CQDs are narrow-bandgap semiconductors that facilitate bandgap alteration from the near-infrared (NIR) to the visible spectrum edge by quantum confinement effects [11–13]. In contrast, HgTe CQDs are characterized by an intrinsic zero bandgap with an inverted band structure and offer tunable bandgaps and optical responses across shortwave infrared (SWIR, 1.5–2.5 μm) [14–16], midwave infrared (MWIR, 3–5 μm) [17–19], and longwave infrared (LWIR, 8–12 μm) [20–22]. Researchers often use device design methodologies, such as vertically stacked structures and planar coupled/patterned devices, to broaden the spectrum range or attain multi-band detection. Vertical stacking improves pixel fill factor and optical coherence. Pejović et al. [23] presented a photodetector based on PbS CQDs and on OPD-PbS CQDs that enables multispectral sensing in a broadband region spanning from visible to SWIR. Zhao et al. [24] demonstrated a visible/shortwave infrared multiband photodetector based on HgTe/CdTe metal-sulfide CQDs with bias-switchable operation. On the other hand, planar patterning offers flexibility for arbitrary band combinations. Tang et al. [25] fabricated a planar three-pixel HgTe CQD multispectral detector using lithography, with each pixel coated with different sized HgTe CQDs covering SWIR, MWIR, and LWIR. Kim et al. [26] developed a monolithic full-color light detector based on PbS and CdSe CQDs. By depositing CQDs with different bandgaps in distinct regions and integrating them with oxide semiconductors, they achieved two-dimensional pixelated multicolor detection across the wavelength range from 365 nm to 1310 nm. In this work, we synthesized PbS CQDs and HgTe CQDs with absorption edges of 1.3 μm, 1.8 μm, 2.6 μm, and 3.5 μm, respectively. Then, the CQDs were used to build planar multiband infrared detectors, which show a distinct photoresponse across the NIR to MWIR. At room temperature, the responsivities are 0.87, 5.39, 2.43, and 0.22 A/W with specific detectivities of 8.87×10^{10}, 2.01×10^{11}, 3.36×10^{10} and 3.54×10^{9} Jones at 1.3 μm, 1.8 μm, 2.6 μm, and 3.5 μm, respectively. Compared with prior studies that use PbS and HgTe CQDs for photodetection [27], the key advantage of our work is the on-chip four-pixel integration using a uniform planar device layout and the same processing flow, enabling spectral coverage from the NIR to the MWIR, as well as co-registered, independently addressable room-temperature multiband detection.

2. Results and Discussions

2.1. Materials Synthesis

In a nitrogen glovebox, four bands of CQDs covering complementary spectral ranges in the infrared were synthesized in this work. These included PbS CQD and three types of HgTe CQDs.

Synthesis (1) Experimental materials: lead (II) oxide (PbO, 99.5%, Sigma-Aldrich, St. Louis, MO, USA), oleic acid (OA, 90%, Sigma-Aldrich), 1-octadecene (ODE, 90%, Sigma-Aldrich), tetramethylsilane (TMS, 99%, Sigma-Aldrich, St. Louis, MO, USA), mercury (II) chloride ($HgCl_2$, >99.5%, Sigma-Aldrich, St. Louis, MO, USA), oleylamine (OAM, 80–90%, Aladdin, Riverside, CA, USA), tellurium powder (Te, 99.999%, Sigma-Aldrich), tri-n-octylphosphine (TOP, 97%, Sigma-Aldrich), 1-dodecanethiol (DDT, >98%, Sigma-Aldrich), and 1,1,2-trichloroethane (TCE, 98%, Aladdin), Didodecyldimethylammonium bromide (DDAB, 98%, Sigma-Aldrich), octane (>99%, Sigma-Aldrich), chlorobenzene (CBZ, 99.5%, Aladdin), methanol (MeOH, 99.8%, Adamas-beta, Shanghai, China), anhydrous ethanol (EtOH, 99.7%, Innochem, Shanghai, China), isopropanol (IPA, Single Metal < 10 ppb, Adamas, Maasdijk, The Netherlands), acetone (>98%).

(2) Synthesis of PbS CQD: A lead precursor solution was prepared by reacting 0.45 g (2 mmol) of PbO with 20 mL of oleic acid at 150 °C for 1 h. Then, the system was then cooled

https://doi.org/10.3390/photonics13010089

to 120 °C and equilibrated for 30 min. Subsequently, a sulfur precursor solution (210 μL TMS in 10 mL ODE) was quickly injected, and the reaction lasted for 2 min. The crude product was refined by precipitation with a methanol/acetone mixture, then recovered by centrifugation and redispersed in 1 mL of octane for further use.

(3) Synthesis of HgTe CQDs: The emission wavelength of HgTe CQDs was tuned across different spectral bands by controlling the reaction temperature and time during a hot-injection synthesis between mercury precursors. The HgTe CQDs synthesis was as follows:

Precursor Preparation: The mercury precursor solution (($HgCl_2$/OAM)) was prepared by dissolving 108.8 mg (0.4 mmol) of $HgCl_2$ in 16 mL of OAM at 100 °C for 1 h. The tellurium precursor solution (TOPTe) was obtained by dissolving 12.76 mg (0.1 mmol) of Te in 0.1 mL of TOP under stirring at 80 °C for 12 h.

Nucleation and Growth: Nucleation was initiated by the rapid injection of 0.4 mL of the TOPTe solution into the thermally equilibrated mercury precursor. This was followed by a growth period at the specified temperature to attain the target bandgap.

Reaction Termination and Purification: The reaction was terminated by the immediate addition of a pre-cooled quenching solution (1.2 mL TOP, 16 mL TCE, 3 mL DDT), followed by rapid cooling in a water bath. Subsequently, the crude CQDs were purified by precipitation with the DDAB/IPA mixture, collected via centrifugation, and finally redispersed in 1 mL of CBZ. A schematic of the synthesis processes for both PbS and HgTe CQDs is presented in Figure 1a.

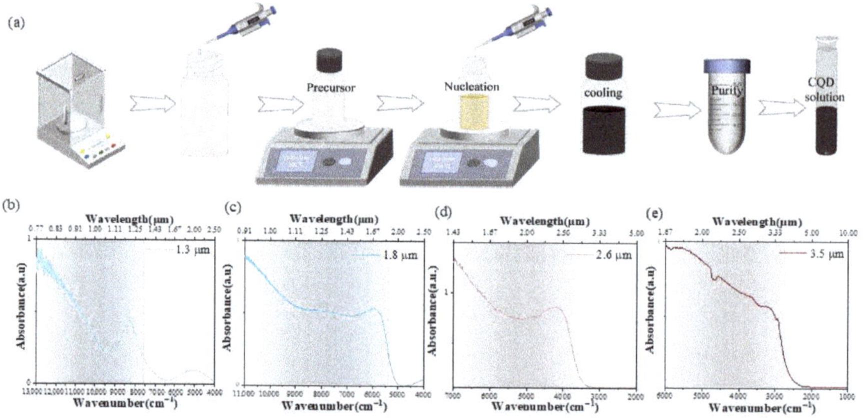

Figure 1. Synthesis and characterization of PbS and HgTe CQDs. (**a**) Schematic illustration of the hot-injection synthesis method. (**b–e**) Absorption spectra of the synthesized CQDs: (**b**) 1.3 μm PbS CQD, (**c**) 1.8 μm HgTe CQD, (**d**) 2.6 μm HgTe CQD, and (**e**) 3.5 μm HgTe CQD.

Absorption Characterization: The optical absorption of the synthesized CQDs was characterized using Fourier transform infrared spectroscopy (FTIR). The cutoff wavelength, defined at 50 percent attenuation on the absorption spectra edge, was determined to be 1.3 μm (7690 cm^{-1}) for PbS CQD, as shown in Figure 1b.

Following the LaMer nucleation-growth model, both reaction temperature and growth time were tuned because they control different stages of CQD formation: temperature governs monomer generation/supersaturation and thus nucleation density and growth kinetics, while time determines the extent of post-nucleation growth before quenching. Spectral tuning was achieved by precisely controlling the reaction parameters during the hot-injection synthesis of HgTe CQDs. The cutoff absorption was tuned from 1.8 μm (5347 cm^{-1}) to 2.6 μm (3850 cm^{-1}) and 3.5 μm (2850 cm^{-1}) by increasing the reaction

 https://doi.org/10.3390/photonics13010089

temperature from 60 °C to 80 °C and 90 °C with growth times of 5 min, 4 min and 6 min, respectively, as shown in Figure 1c–e. Higher synthesis temperatures and adjusted growth times were employed to obtain CQDs with progressively longer cutoff wavelengths. Synthesis parameters of all CQDs are summarized in Table 1.

Table 1. Synthesis parameters for all CQDs targeting different infrared bands.

Wavelength (μm)	Reaction Temperature (°C)	Growth Time (min)
1.3	120	2
1.8	60	5
2.6	80	4
3.5	90	6

2.2. Photodetector Structure and Fabrication

Four pixels with photoactive materials (1.3 μm PbS CQD, 1.8 μm HgTe CQD, 2.6 μm HgTe CQD, and 3.5 μm HgTe CQD) are connected onto a single-plane substrate. Figure 2a,b illustrate the physical configurations and pixel layouts of the four-band photoconductors. Every pixel shares an identical architecture: Au interdigitated electrodes are precisely patterned onto an infrared-transparent Al_2O_3 substrate, which functions as the device's foundational platform. Subsequently, a continuous layer of CQDs is deposited over this substrate, ensuring complete coverage and electrical connection between the Au contacts. This CQD film serves as the active medium for infrared detection, while the Au electrodes efficiently collect the photogenerated charge carriers, as depicted in the cross-sectional schematic of a single pixel shown in Figure 2c. The preparation processes for the PbS and HgTe pixels are detailed below.

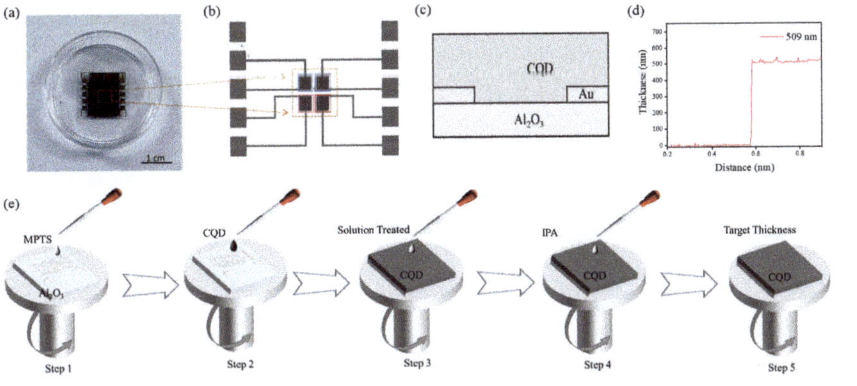

Figure 2. Structure and fabrication of the multiband device. (**a**) Photograph of the fabricated device. (**b**) Pixel arrangement of the four-band photoconductors. (**c**) Schematic cross-section of a single pixel. (**d**) The thickness characterization for the CQD films. (**e**) Fabrication process of multiband device.

(1) Substrate Pretreatment:

The Al_2O_3 substrate was first treated with (3-mercaptopropyl) trimethoxysilane (MPTS) for 1 min to increase film adhesion, then rinsed with isopropanol (IPA) and spin-dried at 3000 rpm for 15 s.

https://doi.org/10.3390/photonics13010089

(2) Film Deposition and Processing

For the pixel with PbS CQDs, the PbS CQDs ink was spin-coated at 2000 rpm for 20 s. Ligand exchange was carried out by immersing the film in a methanolic solution of tetrabutylammonium iodide (TBAI, 10 mg/mL) for 30 s.

For the three pixels with HgTe CQDs, the CQDs' ink was spin-coated at 1500 rpm for 15 s. Ligand exchange was accomplished by immersing the film in a solution of EDT, HCl, and IPA (volume ratio 1:1:50) for 10 s.

(3) Layer Completion and Thickness Control

Following ligand exchange, the film was rinsed with IPA to remove excess ligands and spin-dried at 2000 rpm for 10 s. Steps 2–4 (spin-coating, solution treatment, and rinsing/drying) were repeated cyclically until the target film thickness of approximately 500 nm was achieved, the thickness characterization for the CQD films using a step meter, as shown in Figure 2d. The schematic diagram of the device fabrication process as shown in Figure 2e.

This integrated structure and fabrication approach facilitates the monolithic production of four-band devices, thereby enabling simultaneous multi-spectral detection. It lays a fundamental device platform for subsequent spectral recognition and extraction of multidimensional information.

2.3. Photodetector Characterization

(1) Responsivity (R) and Detectivity (D^*)

The responsivity (R) and specific detectivity (D^*) were used to evaluate device performance. R is a parameter that describes detector photoelectric conversion capability by defining the relationship between the input power and the output signal of the detector.

$$R = \frac{I_{\mathrm{ph}}}{P} \tag{1}$$

where I_{ph} is the photocurrent of the detector, P is the input power.

The D^* is the key parameter for characterizing the sensitivity of a photodetector. It enables a direct comparison of the detection capabilities among devices with different specifications, as its normalization process eliminates the effects of detector size and noise bandwidth on performance evaluation. A higher D^* value indicates a stronger capability of the detector to detect weak light signals.

$$D^* = \frac{\sqrt{A_d \Delta f}}{i_n} R \tag{2}$$

where R is responsivity, i_n is noise current density, A_d is the effective illuminated area, and Δf is the measurement bandwidth.

The devices were measured at room temperature, and their I-V curves were systematically monitored using a Keithley 2602B (Keithley Instruments, Solon, OH, USA)source meter for bias application and current collection, as shown in Figure 3a–d. The 1.3 μm PbS CQD pixel was characterized using a tungsten–halogen lamp (Silan, Hangzhou, China) as the near-infrared light source, whereas the 1.8 μm, 2.6 μm, and 3.5 μm HgTe CQD pixels were examined using a blackbody of 600 °C as the light source.

 https://doi.org/10.3390/photonics13010089

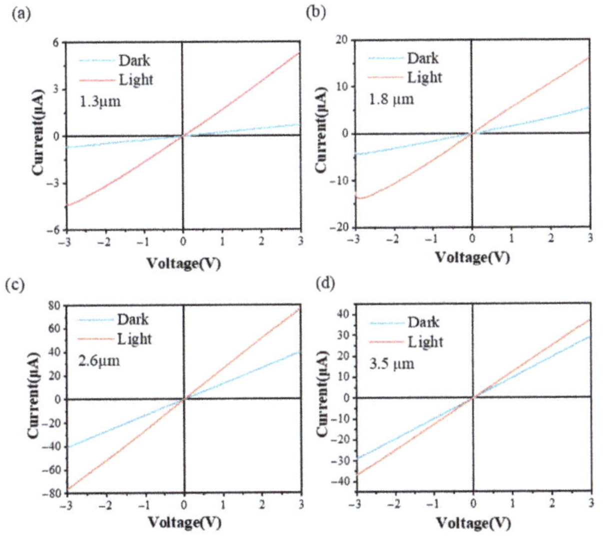

Figure 3. Current–Voltage (I-V) characteristics of the four pixels measured under dark and illuminated conditions: (**a**) 1.3 μm PbS CQD pixel, (**b**) 1.8 μm HgTe CQD pixel, (**c**) 2.6 μm HgTe CQD pixel, and (**d**) 3.5 μm HgTe CQD pixel.

The multiband device operates in the photoconductive mode. For incident photons with energies exceeding the bandgap of the CQDs in a given pixel, the active layer efficiently absorbs the light and generates electron–hole pairs. Under an applied bias, these photogenerated carriers are separated and transported through the percolated CQDs, leading to an increase in conductivity compared to the dark state. The resulting photocurrent depends on the incident photon flux and the carrier transport properties in the film, so that each pixel selectively responds within its designed spectral band. By integrating CQDs with different bandgaps in spatially separated pixels, the device achieves multiband photoconductive detection from the NIR to the MWIR on a single planar platform. Table 2 summarizes the performance metrics of every pixel measured under the 3 V bias. A photocurrent of 4.61 μA is produced for the 1.3 μm PbS CQD pixel with an incident optical power of 5.30 μW. This results in a responsivity of 0.87 A/W and a specific detectivity of 8.87×10^{10} Jones. The 1.8 μm HgTe CQD pixel exhibits the best performance with a photocurrent of 10.67 μA at 1.98 μW, a high responsivity of 5.39 A/W, and a detectivity of 2.01×10^{11} Jones. The photocurrents at longer wavelengths are 36.91 μA (15.14 μW) and 8.41 μA (38.51 μW) for the 2.6 μm and 3.5 μm pixels, the corresponding responsivities are 2.43 and 0.22 A/W, while the detectivities are 3.36×10^{10} and 3.54×10^{9} Jones, respectively. This trend is mainly attributed to the reduced bandgap at longer cutoff wavelengths, which increases thermally activated carriers and dark current at room temperature, thereby lowering the signal-to-noise ratio and detectivity.

(2) Photoresponse

The device being tested was utilized as an external detector for FTIR. The incident optical signals were converted into corresponding photocurrents by the detector. Subsequently, the output was then fed into a Femto DLPCA-200 current amplifier(Femto, Berlin Germany), and then into a Femto DLPVA-100-B voltage amplifier(Femto, Berlin Germany) for external signal conditioning, before being routed to the internal detector port of the FTIR. System control was managed through dedicated electronic modules and software, facilitating fully automated operation and the acquisition of response spectra

https://doi.org/10.3390/photonics13010089

for four devices, as illustrated in Figure 4a–d. The cut-off wavelength is defined as the wavelength at which the spectral responsivity falls to half of its peak value. According to the photoresponse spectra, the PbS CQD pixel and the three HgTe CQDs pixels have cut-off wavelengths of 7532 cm^{-1} (1.3 μm), 5344 cm^{-1} (1.8 μm), 3808 cm^{-1} (2.6 μm), and 2847 cm^{-1} (3.5 μm), respectively. These results are consistent with the absorption spectrum characteristics of the relevant materials, suggesting that the spectral response range of the devices is primarily determined by the inherent absorption capabilities of the CQD. Minor differences between the absorption and photoresponse spectra are expected due to device-related effects (e.g., film thickness, electrodes/substrate reflection and interference) and the DTGS-normalized FTIR measurement.

Table 2. The performance parameters of all pixels.

Pixel	P (μW)	I_{ph} (μA)	R (A/W)	D^* (Jones)
1.3 μm PbS CQD	5.30	4.61	0.87	8.87×10^{10}
1.8 μm HgTe CQD	1.98	10.67	5.39	2.01×10^{11}
2.6 μm HgTe CQD	15.14	36.91	2.43	3.36×10^{10}
3.5 μm HgTe CQD	38.51	8.41	0.22	3.54×10^{9}

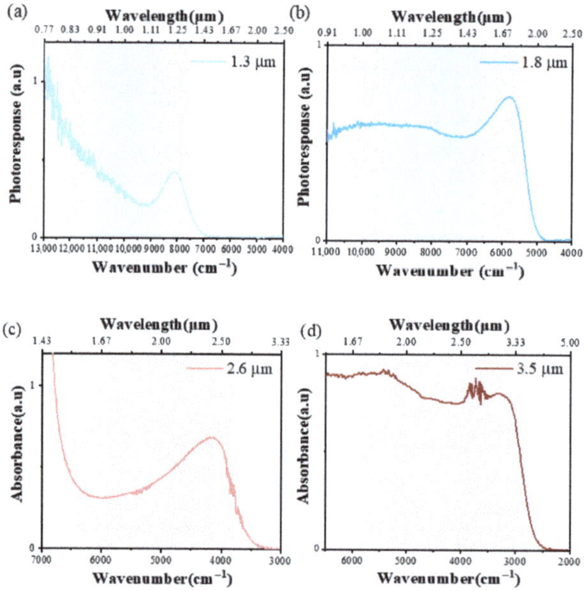

Figure 4. Room-temperature spectral response of the four pixels, normalized to the DTGS: (**a**) 1.3 μm PbS CQD pixel, (**b**) 1.8 μm HgTe CQD pixel, (**c**) 2.6 μm HgTe CQD pixel, and (**d**) 3.5 μm HgTe CQD pixel.

For comparison with the CQD-based multispectral photodetectors, representative literature results are summarized in Table 3, highlighting the distinct integration strategies and performance trade-offs across different multiband platforms.

https://doi.org/10.3390/photonics13010089

Table 3. Performance comparison of representative CQD-based multiband photodetectors.

Materials	Device Structure	Operating Temperature	Bias (V)	Wavelength (μm)	Detectivity (Jones)	Responsivity (A/W)	Ref
PbS CQDs/OPD-PbS CQDs	Vertical stacking	/	1	0.4–1.0 0.8–1.2	/ /	EQE = 70% @ 500nm EQE = 30% @ 1150nm	[23]
HgTe/CdTe CQDs	Vertical stacking	room tempreture	+3 −2	0.7 2.1	1.1×10^{11} 4.5×10^{11}	0.5 1.1	[24]
HgTe CQD	Planar patterning	room tempreture	<10	4.8 6.0 9.5	2×10^{7}@2 μm 1.25×10^{7}@4 μm 1.0×10^{7}@7 μm	0.1@2 μm 0.07@4 μm 0.05@7 μm	[25]
PbS/CdSe/CdS CQD	Planar patterning	room tempreture	VD = 15 VG = −3	406 530 630 1310	4.2×10^{17}	8.3×10^{3}	[26]
PbS/HgTe CQD	Planar patterning	room tempreture	3	1.3 1.8 2.6 3.5	8.87×10^{10} 2.01×10^{11} 3.36×10^{10} 3.54×10^{9}	0.87 5.39 2.43 0.22	This work

3. Conclusions

In this work, we fabricated a multiband planar integrated photodetector which consists of four pixels. Each pixel is based on different photoactive materials corresponding to four specific bands: 1.3 μm PbS CQD, 1.8 μm HgTe CQD, 2.6 μm HgTe CQD, and 3.5 μm HgTe CQD. This design allows for comprehensive spectral coverage spanning from the NIR to the MWIR range. All pixels demonstrated good photoresponse characteristics, with the 1.8 μm HgTe CQD pixel exhibiting the highest performance at room temperature, with the responsivity of 5.39 A/W and the specific detectivity of 2.01×10^{11} Jones at 3 V bias voltage. The integration of multiple detection bands onto a single compact chip offers an efficient platform for simultaneous multispectral imaging, enabling advanced functionalities such as precise spectral fingerprinting and detailed multidimensional scene analysis.

Author Contributions: Y.X., X.X. and M.C. designed the experiments; Y.X. and X.X. conceived and designed the idea; Y.X. and X.X. collected and analyzed the data. Y.X. and X.X. wrote the manuscript. M.C. revised the manuscript. Y.X., X.X., L.W., Z.G., Q.H. and M.C. reviewed this work. All authors have read and agreed to the published version of the manuscript.

Funding: This work is financially supported by National Natural Science Foundation of China (No.62475012), State Key Laboratory of Laser Interaction with Matter (SKLLIM-G-2501), National Key Laboratory of Infrared Detection Technologies (IRDT-ZGKXY-25-04), Natural Science Foundation of Zhejiang Province (LD25F040001).

Data Availability Statement: The original contributions presented in this study are included in the article. Further inquiries can be directed to the corresponding authors.

Conflicts of Interest: The authors declare no conflict of interest.

References

1. Rogalski, A.; Antoszewski, J.; Faraone, L. Third-generation infrared photodetector arrays. *J. Appl. Phys.* **2009**, *105*, 091101. [CrossRef]
2. Shen, Y.; Li, J.; Lin, W.; Chen, L.; Huang, F.; Wang, S. Camouflaged Target Detection Based on Snapshot Multispectral Imaging. *Remote Sens.* **2021**, *13*, 3949. [CrossRef]
3. Cumis, M.S.; Viciani, S.; Borri, S.; Patimisco, P.; Sampaolo, A.; Scamarcio, G.; De Natale, P.; D'Amato, F.; Spagnolo, V. Widely-tunable mid-infrared fiber-coupled quartz-enhanced photoacoustic sensor for environmental monitoring. *Opt. Express* **2014**, *22*, 28222–28231. [CrossRef] [PubMed]

 https://doi.org/10.3390/photonics13010089

4. Thompson, N.A. Common Aperture Multi-Spectral Optics for Military Applications. *Proc. SPIE* **2012**, *8353*, 83531X.
5. Thompson, N.A. Optical Design of Common Aperture, Common Focal Plane, Multispectral Optics for Military Applications. *Opt. Eng.* **2013**, *52*, 061308. [CrossRef]
6. Wang, F.; Zhong, Y.; Bruns, O.; Liang, Y.; Dai, H. In vivo NIR-II Fluorescence Imaging for Biology and Medicine. *Nat. Photonics* **2024**, *18*, 535–547. [CrossRef]
7. Kopytko, M.; Gawron, W.; Kębłowski, A.; Stępień, D.; Martyniuk, P.; Jóźwikowski, K. Numerical analysis of HgCdTe dual-band infrared detector. *Opt. Quantum Electron.* **2019**, *51*, 62. [CrossRef]
8. Kaniewski, J.; Muszalski, J.; Piotrowski, J. Recent advances in InGaAs detector technology. *Phys. Status Solidi (A)* **2004**, *201*, 2281–2287. [CrossRef]
9. Ekimov, A.; Efros, A.L.; Onushchenko, A.A. Quantum size effect in semiconductor microcrystals. *Solid State Commun.* **1993**, *88*, 947–950. [CrossRef]
10. Garcia de Arquer, F.P.; Talapin, D.V.; Klimov, V.I.; Arakawa, Y.; Bayer, M.; Sargent, E.H. Semiconductor quantum dots: Technological progress and future challenges. *Science* **2021**, *373*, eaaz8541. [CrossRef]
11. Balazs, D.M.; Nugraha, M.I.; Bisri, S.Z.; Sytnyk, M.; Heiss, W.; Loi, M.A. Reducing charge trapping in PbS colloidal quantum dot solids. *Appl. Phys. Lett.* **2014**, *104*, 112104. [CrossRef]
12. Wang, C.; Wang, Y.; Jia, Y.; Wang, H.; Li, X.; Liu, S.; Liu, X.; Zhu, H.; Wang, H.; Liu, Y.; et al. Precursor Chemistry Enables the Surface Ligand Control of PbS Quantum Dots for Efficient Photovoltaics. *Adv. Sci.* **2022**, *10*, 2204655. [CrossRef] [PubMed]
13. Lee, J.W.; Kim, D.Y.; Baek, S.; Yu, H.; So, F. Inorganic UV-Visible-SWIR Broadband Photodetector Based on Monodisperse PbS Nanocrystals. *Small* **2016**, *12*, 1328–1333. [CrossRef] [PubMed]
14. Yang, J.; Hu, H.; Lv, Y.; Yuan, M.; Wang, B.; He, Z.; Chen, S.; Wang, Y.; Hu, Z.; Yu, M.; et al. Ligand-Engineered HgTe Colloidal Quantum Dot Solids for Infrared Photodetectors. *Nano Lett.* **2022**, *22*, 3465–3472. [CrossRef]
15. Xue, X.; Luo, Y.; Hao, Q.; Cao, J.; Tang, X.; Liu, Y.; Chen, M. Low Dark-Current Quantum-Dot Infrared Imager. *ACS Photonics* **2023**, *10*, 4290–4298. [CrossRef]
16. Sergeeva, K.A.; Hu, S.; Sokolova, A.V.; Portniagin, A.S.; Chen, D.; Kershaw, S.V.; Rogach, A.L. Obviating Ligand Exchange Preserves the Intact Surface of HgTe Colloidal Quantum Dots and Enhances Performance of Short Wavelength Infrared Photodetectors. *Adv. Mater.* **2024**, *36*, e2306518. [CrossRef]
17. Chen, M.; Lan, X.; Tang, X.; Wang, Y.; Hudson, M.H.; Talapin, D.V.; Guyot-Sionnest, P. High Carrier Mobility in HgTe Quantum Dot Solids Improves Mid-IR Photodetectors. *ACS Photonics* **2019**, *6*, 2358–2365. [CrossRef]
18. Xue, X.; Chen, M.; Luo, Y.; Qin, T.; Tang, X.; Hao, Q. High-operating-temperature mid-infrared photodetectors via quantum dot gradient homojunction. *Light Sci. Appl.* **2023**, *12*, 2. [CrossRef]
19. Caillas, A.; Guyot-Sionnest, P. Uncooled high detectivity mid-infrared photoconductor using HgTe quantum dots and nanoantennas. *ACS Nano* **2024**, *18*, 8952–8960. [CrossRef]
20. Keuleyan, S.E.; Guyot-Sionnest, P.; Delerue, C.; Allan, G. Mercury telluride colloidal quantum dots: Electronic structure, size-dependent spectra, and photocurrent detection up to 12 μm. *ACS Nano* **2014**, *8*, 8676–8682. [CrossRef]
21. Zhang, H.; Peterson, J.C.; Guyot-Sionnest, P. Intraband Transition of HgTe Nanocrystals for Long-Wave Infrared Detection at 12 μm. *ACS Nano* **2023**, *17*, 7530–7538. [CrossRef] [PubMed]
22. Xue, X.; Hao, Q.; Chen, M. Very long wave infrared quantum dot photodetector up to 18 μm. *Light Sci. Appl.* **2024**, *13*, 89. [CrossRef] [PubMed]
23. Pejović, V.; Georgitzikis, E.; Lieberman, I.; Malinowski, P.E.; Heremans, P.; Cheyns, D. Photodetectors Based on Lead Sulfide Quantum Dot and Organic Absorbers for Multispectral Sensing in the Visible to Short-Wave Infrared Range. *Adv. Funct. Mater.* **2022**, *32*, 2201424. [CrossRef]
24. Zhao, P.; Qin, T.; Mu, G.; Zhang, S.; Luo, Y.; Chen, M.; Tang, X. Band-engineered dual-band visible and short-wave infrared photodetector with metal chalcogenide colloidal quantum dots. *J. Mater. Chem. C* **2023**, *11*, 2842–2850. [CrossRef]
25. Tang, X.; Lai, K.W.C. Scalable Fabrication of Infrared Detectors with Multispectral Photoresponse Based on Patterned Colloidal Quantum Dot Films. *ACS Photonics* **2016**, *3*, 2396–2404. [CrossRef]
26. Kim, J.; Kwon, S.M.; Kang, Y.K.; Kim, Y.; Lee, M.J.; Han, K.; Facchetti, A.; Kim, M.; Park, S.K. A skin-like two-dimensionally pixelized full-color quantum dot photodetector. *Sci. Adv.* **2019**, *5*, eaax8801. [CrossRef]
27. Hu, C.; Gassenq, A.; Justo, Y.; Yakunin, S.; Heiss, W.; Hens, Z.; Roelkens, G. Short-wave infrared colloidal quantum dot photodetectors on silicon. *Proc. SPIE* **2013**, *8631*, 863127.

 https://doi.org/10.3390/photonics13010089

photonics

MDPI

Article

Fused Complementary 3D Reconstruction Based on Polarization Binocular Line-Structured Light

Mingsheng Liu [1], Hongyuan Zhou [1], Sisheng Nie [2], Yan Jiang [1], Zhong Wu [3], Dahai Xu [3], Ling Zhu [4], Yanliang Zhan [2] and Zhenmin Zhu [2,*]

1 Jiangxi Traffic Monitoring Command Center, Nanchang 330036, China
2 School of Electrical and Electronic Engineering, East China Jiaotong University, Nanchang 330013, China
3 Jiangxi Fangxing Technology Co., Ltd., Nanchang 330025, China
4 Jiangxi Communications Design and Research Institute Co., Ltd., Nanchang 330001, China
* Correspondence: 2707@ecjtu.edu.cn

Abstract

Line-structured light three-dimensional (3D) measurement is commonly used for three-dimensional contour reconstruction of objects in complex industrial environments, but the problem of missing information occurs when three-dimensional reconstruction is performed on objects with smooth surfaces, single texture, and high reflectivity, resulting in defective reconstructed object surfaces. For this reason, this study proposes a fused complementary 3D reconstruction technique based on a polarization-based binocular line-structured light system. First, the reconstructed image of the object is captured using a Polarization Binocular Camera, and the polarized imaging effectively reduces the strong highlights and extracts more detailed information on the surface of the object. Then, the calibrated camera and optical planes are used to acquire the spatial coordinates of the object reconstructed by the left camera and right camera. Finally, the spatial coordinates obtained by the left camera and right camera are aligned, and the high-precision 3D reconstruction results are generated. The experimental results show that the proposed method can effectively improve the accuracy and robustness of 3D reconstruction, has a good application prospect, and can meet the technical requirements of industrial 3D measurement.

Keywords: polarization line-structured light; binocular; 3D reconstruction

1. Introduction

Three-dimensional measurement based on multi-view imaging has a long history in photogrammetry and computer vision. As early as the 1980s, automated stereo matching techniques were developed to reconstruct 3D point clouds from overlapping images captured by two or more cameras from different viewpoints [1]. A landmark survey by Mayer in 1988, published under the auspices of the International Society for Photogrammetry and Remote Sensing (ISPRS), provided a comprehensive review of stereo matching algorithms available at the time, many of which laid the foundation for modern 3D reconstruction pipelines [2]. The theoretical and practical framework of digital photogrammetry—formalized over decades by researchers such as Kraus [3], Luhmann et al. [4], and McGlone [5]—has established rigorous methodologies for geometric reconstruction from imagery, forming a cornerstone of non-contact 3D measurement.

 https://doi.org/10.3390/photonics13030238

In recent years, advances in structured light projection, high-resolution sensors, and real-time image processing have further enhanced the accuracy, robustness, and applicability of 3D measurement systems—particularly in industrial inspection, reverse engineering, and dynamic scene capture.

In many application fields, such as virtual reality [6], augmented reality [7], industrial measurement [8], etc., accurate and fast acquisition and reconstruction of real-world 3D information has become an important task. Line-structured light 3D reconstruction, as a common 3D reconstruction method, can obtain the 3D shape information of the measured object by projecting light and analyzing its deformation information. It has the advantages of non-contact, fast imaging speed, high measurement accuracy, and a strong sense of reality. It has been widely used in industrial environments [9–11].

In previous decades, various 3D reconstruction methods have been developed to meet the needs in different application scenarios. The traditional 3D reconstruction methods are mainly based on the principles of visual geometry, which mainly include techniques such as stereo vision [12] and structured light [13–16]. Pioneering work by Faugeras [17] and Hartley & Zisserman [18] formalized the geometric foundations of multi-view reconstruction, while Salvi et al. [19] and Geng [20] provided comprehensive surveys of structured light techniques that remain widely cited today. The line-structured light 3D reconstruction method belongs to one kind of structured light technology, which utilizes the principle of light projection and reflection to obtain the depth information on the surface of the measured object. The basic principle of line-structured light 3D reconstruction is that a series of light lines is projected on the surface of the object to be measured, and the deformation of the light lines on the surface of the object is analyzed. The camera acquires the contour image, detects the 2D image coordinates of the structured light stripe, and calculates the corresponding 3D coordinates using the proposed algorithm. The essence of the process of online-structured light 3D reconstruction is how to convert the two-dimensional light stripe coordinate information into three-dimensional coordinate information to infer the depth information of the object surface. Many studies have investigated the line-structured light 3D reconstruction method. For example, Hao et al. [21] proposed a 3D reconstruction technique for pavements based on line-structured light. Chen et al. [22] proposed a monocular line-structured light-based 3D scanning method for large workpieces. A combined monocular and binocular method for scanning line-structured light was used by Wang et al. [23]. The scanning is performed in a binocular system, and iterative eigenvalue least squares are used to fit the 3D point coordinates of the center of the acquired light stripes to obtain the light plane equation. Finally, the obtained light plane equation is utilized in the monocular system to calculate the 3D point coordinates of the object surface based on the geometric principle of perspective projection. Guo et al. [24] built a 3D point cloud measurement system based on a line-structured light sensor and a high-precision air-bearing rotary table. It can quickly obtain the 3D shape information of gears and tooth flanks. Finally, the profile and pitch errors were calculated and evaluated. Wang et al. [25] proposed a monocular camera 3D reconstruction method based on an improved line-structured light marker localization method and the Scheimpflug principle. These structured light 3D reconstruction methods use a scanning system consisting of a camera and a line-structured light emitter. This method is simple in composition and can scan objects quickly. However, some problems are encountered when scanning the surface of complex objects. Occlusions and highlights on the surface of the object itself can cause some information to be lost during the 3D reconstruction process. This means that the 3D shape of the object cannot be recovered accurately and completely. To overcome this occlusion limitation inherent in conventional structured light systems, we integrate a laser line projector with a dual-camera stereo setup. This hybrid configuration leverages geometric redundancy: even

 https://doi.org/10.3390/photonics13030238

if a surface point is occluded in one camera view, the laser stripe observed by the other camera—combined with the calibrated projector–camera geometry—enables triangulation of the otherwise invisible region.

The structure of this paper is as follows: the first part is the introduction; the second part introduces the line-structured optical vision sensor and the dual-camera model, as well as the transformation between each coordinate system; the third part is the principle of the 3D measurement system; the fourth part is the experimental part, and the experimental results validate the feasibility of the method; and the fifth part is the conclusion.

2. Basic Principles and Models

2.1. Model of the Line-Structured Light Vision Sensor

Aiming at the above problems, the work in this paper is inspired by the line-structured light scanning method based on the combination of monocular and binocular proposed by Wang and the reconstruction method by Hu [26] using an integrated measurement system with multiple line-structured light sensors. A complementary 3D reconstruction technique based on polarized binocular line-structured light fusion is proposed. To obtain a 3D model of the object with richer 3D point cloud data and higher accuracy by one scan. A rotary scanning 3D measurement system was constructed using two CCD cameras with polarizers in front and a laser transmitter. The left and right cameras were combined with the laser to form two measurement systems, i.e., the left camera and the right camera, respectively. And the 3D model of the object with richer point cloud information is obtained through data fusion. In this paper, firstly, the camera and the optical plane are calibrated for the left camera and the right camera, respectively, to obtain the interior and external parameters of the camera and the optical plane equations. After that, the triangulation principle is utilized to obtain the spatial 3D data of the objects to be measured by the left camera and the right camera, respectively. Taking the coordinate system of the left camera as the world coordinate system, the 3D coordinate data under the corresponding camera coordinate system of the right camera are unified into the world coordinate system through the rotation translation matrix of the relative position relationship between the two cameras. Finally, the rotational translation matrix of the left camera is then used to transform all the 3D camera coordinate points to the world coordinate system for point cloud data fusion.

A line-structured light vision sensor contains two main components: a laser and a CCD camera. As shown in Figure 1. The laser generates a beam of light that forms an optical plane that intersects the object to be measured. This light plane projects on the object a light stripe L reflecting the contour of the object. The points on the light stripe L are the points we want to measure, and they are imaged by the camera onto the image plane. By establishing a perspective correspondence between the light plane and the image plane, we can calculate the coordinates of the corresponding measured points in space from the positions of the image points. In Figure 1, O_c is the perspective projection center of the camera, and the image plane O_c of the camera is the distance to the point f called the effective focal length. $O_w X_w Y_w Z_w$ is the world coordinate system (WCS), $O_c X_c Y_c Z_c$ is the camera coordinate system (CCS), and $O_u X_u Y_u$ is the image coordinate system (ICS). The line-structured light vision sensor utilizes the principle of triangular sectioning to obtain the 3D coordinate information on the light bar.

https://doi.org/10.3390/photonics13030238

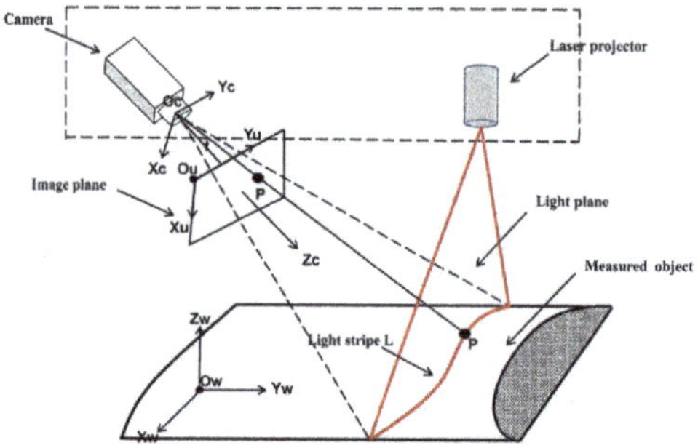

Figure 1. Diagram of the line-structured light 3D measurement model.

An arbitrary point P on the optical plane is assumed to have a projection point p on the image plane, and the point P has the coordinates $P_c(X_c, Y_c, Z_c)^T$ on the camera coordinate system, the point p has the coordinates $P_w(X_w, Y_w, Z_w)^T$ on the world coordinate system, and the corresponding pixel-planar coordinate system is $P(u,v)^T$ on the pixel-planar coordinate system, and the corresponding image coordinate system is $P(X,Y)^T$ on the image coordinate system.

According to the pinhole camera model, the perspective projection from a 3D point in the world coordinate system to its corresponding pixel coordinates in the image plane is given by Equation (1):

$$s\begin{bmatrix} u \\ v \\ 1 \end{bmatrix} = N\begin{bmatrix} R & T \\ \vec{0} & 1 \end{bmatrix}\begin{bmatrix} x_w \\ y_w \\ z_w \\ 1 \end{bmatrix} = M\begin{bmatrix} x_w \\ y_w \\ z_w \\ 1 \end{bmatrix} \quad (1)$$

where s is an arbitrary scale factor, N denotes the interior matrix, and R and T represent the rotation and translation between the world and camera coordinate systems, respectively.

R denotes the 3×3 orthogonal rotation matrix, T denotes the 3×1 translation matrix vector, and $\vec{0}$ denotes the 1×3 zero matrix. M is the projection matrix of the camera. N is the matrix of interior parameters of the camera. Defined as Equation (2):

$$N = \begin{bmatrix} f_x & 0 & u_0 & 0 \\ 0 & f_y & v_0 & 0 \\ 0 & 0 & 1 & 0 \end{bmatrix} \quad (2)$$

where f_x and f_y denote the effective focal lengths in the column and row directions of the image sensor, respectively. These two parameters are introduced to account for the possible non-square geometry of the image pixels, which can cause different scaling factors along the image axes. The point (u_0, v_0) denotes the principal point (optical center).

To correct for lens imperfections, we model both radial and tangential distortions using the standard Brown–Conrady distortion model. The relationship between the distorted pixel coordinates (u_d, v_d) and the ideal undistorted coordinates (u, v) is given by Equations (3) and (4):

$$u_d = u + (u - u_0)\left(k_1 r^2 + k_2 r^4\right) + \left[2p_1(u - u_0)(v - v_0) + p_2\left(r^2 + 2(u - u_0)^2\right]\right. \quad (3)$$

$$u_d = v + (v - v_0)\left(k_1 r^2 + k_2 r^4\right) + \left[2p_2(u - u_0)(v - v_0) + p_1\left(r^2 + 2(v - v_0)^2\right)\right] \quad (4)$$

https://doi.org/10.3390/photonics13030238

where $r^2 = (u - u_0)^2 + (v - v_0)^2$, and k_1, k_2 are radial distortion coefficients, while p_1, p_2 are tangential distortion coefficients.

In this paper, all interior parameters—including $f_x, f_y, u_0, v_0, k_1, k_2, p_1$ and p_2—are calibrated using Zhang's method [27], extended with an optimal polarization angle strategy [28], and implemented via the MATLAB Camera Calibrator toolbox (Version R2022a, The MathWorks, Inc., Natick, MA, USA) [29]. The p point satisfies the optical plane equation, and the line-structured light vision sensor utilizes the triangulation principle to obtain 3D coordinate information on the light stripe. The projected light plane equation can be expressed as Equation (5):

$$Z_W = Ax_W + By_W + C \tag{5}$$

The parameters A, B, and C in the light plane equation (Equation (3)) are determined through a calibration procedure using the checkerboard pattern mounted on the rotating platform. Specifically, the laser stripe is projected onto the checkerboard at multiple rotational angles, and the corresponding 3D coordinates of points on the stripe are reconstructed using the calibrated camera model. These 3D points are then fitted to a plane via least-squares regression to obtain A, B, and C.

The model parameters of the line-structured light sensor include the interior parameter matrix N of the camera and the optical plane equations. If the above parameters are known, the three-dimensional coordinates of the point on the optical stripe can be obtained from the pixel coordinates of the point on the camera plane using Equations (1)–(5).

2.2. *Calibration of the Light Plane of Line-Structured Light*

The commonly used methods for the calibration of light plane parameters are the filament scattering method [30], sawtooth target method [31], and cross-ratio invariant method [32]. In this paper, based on the existing chessboard in the laboratory as a plane target, the cross-ratio invariant method based on the free moving plane target is used to solve the parameters of the optical plane equation by extracting the feature points and using the nonlinear least squares method. The initial estimates for the plane parameters (a, b, c, d) in the nonlinear least squares optimization are obtained from a preliminary linear least squares fit to the same 3D calibration points. Specifically, we first solve the linear system $Ax = 0$ (subject to $||x|| = 1$) via singular value decomposition (SVD), yielding a robust initial plane normal and offset. These values serve as the starting point for the subsequent nonlinear refinement, ensuring rapid convergence and avoiding local minima.

Before carrying out the calibration of the optical plane of the line structure, the camera must first be well calibrated, and the optical plane is calibrated according to the parameters after the calibration is completed. The specific steps of optical plane calibration are as follows:

Step 1: Keep the relative position of the camera and the laser unchanged and adjust the aperture size. A checkerboard target is placed at least twice at suitable locations within the camera's field of view as data for parameter calibration. The light streaks are projected onto the checkerboard target by the line laser.

Step 2: The camera captures images of the checkerboard pattern with light streaks, extracts the 2D coordinates of the feature points on each image, and calculates the intersection ratio value in the 2D plane to find the 3D coordinates of the unknown feature points based on the intersection ratio invariant method.

Step 3: The coordinate system conversion is accomplished by converting the acquired images to the camera coordinate system and then to the world coordinate system through the target-camera conversion matrix.

Step 4: After the optical plane parameters are determined, the corresponding spatial 3D coordinates (X_w, Y_w, Z_w) can be solved according to the interior parameters of the

https://doi.org/10.3390/photonics13030238

camera and the optical plane equations, after the center coordinates $(x_0,\ y_0)$ are extracted from the light bar image acquired each time.

2.3. Dual-Camera Model and Conversion Between Camera Coordinate Systems

As shown in Figure 2, the system is a two-camera line-structured light measurement setup, which consists of two cameras to reduce occlusion and enable more comprehensive 3D reconstruction. The left camera coordinate system is $O_L X_L Y_L Z_L$, defined with its origin at the optical center of the left camera, and the right camera coordinate system $O_R X_R Y_R Z_R$ is similarly defined for the right camera. The world coordinate system $O_W X_W Y_W Z_W$ originates at the reference point of the reconstructed object, as established during calibration.

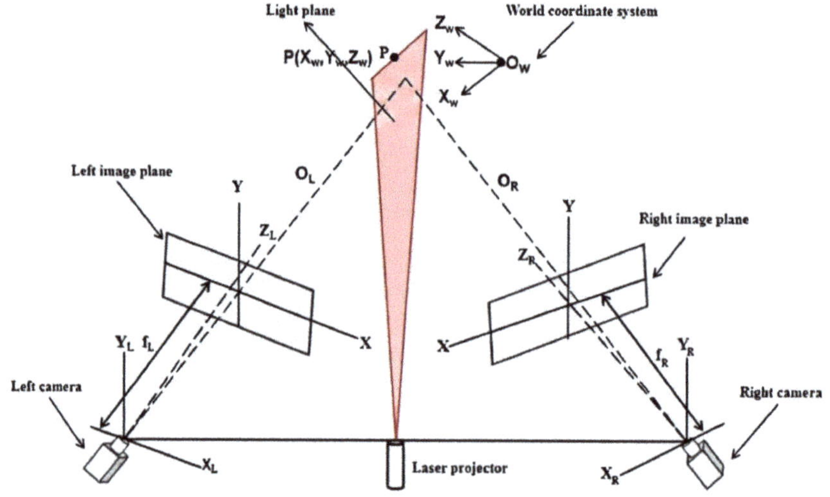

Figure 2. Dual camera model.

Each camera possesses two effective focal lengths corresponding to the row and column pixel dimensions, denoted as f_{xL}, f_{yL} for the left camera and f_{xR}, f_{xR} for the right camera. In Figure 2, for simplicity of illustration, the symbols f_L and f_R are used to represent the row focal lengths f_{xL} and f_{xR}, respectively.

Let an arbitrary point on an object projected by a laser stripe have world coordinates $P(X_w, Y_w, Z_w)$. Its coordinates under the left camera coordinate system are $P_L(X_L, Y_L, Z_L)$. The transformation between the world coordinate system and the left and right camera coordinate systems is expressed as Equations (6) and (7):

$$\begin{bmatrix} X_L \\ Y_L \\ Z_L \end{bmatrix} = R_L \times \begin{bmatrix} X_W \\ Y_W \\ Z_W \end{bmatrix} + T_L \tag{6}$$

$$\begin{bmatrix} X_R \\ Y_R \\ Z_R \end{bmatrix} = R_R \times \begin{bmatrix} X_W \\ Y_W \\ Z_W \end{bmatrix} + T_R \tag{7}$$

https://doi.org/10.3390/photonics13030238

The coordinate transformation relationship between the left and right cameras is given by Equation (8):

$$\begin{bmatrix} X_R \\ Y_R \\ Z_R \end{bmatrix} = R_{LR} \times \begin{bmatrix} X_L \\ Y_L \\ Z_L \end{bmatrix} + T_{LR} \tag{8}$$

In Equations (6)–(8), R_L, R_R, and R_{LR} denote rotation matrices, while T_L, T_R, and T_{LR} denote translation vectors. These parameters are obtained through extrinsic calibration using the MATLAB Camera Calibration Toolbox. Although the intrinsic parameters of the two cameras differ slightly (as shown in Tables 1 and 2), the extrinsic calibration process—based on a high-precision checkerboard observed from multiple views—implicitly accounts for these differences by optimizing R_{LR} and T_{LR} in the metric space defined by the calibration object. Therefore, no explicit scale factor is required for point cloud fusion, as validated by the sub-millimeter reconstruction accuracy in Section 4.

Table 1. Interior parameters of the left camera.

Physical Meaning	Parameters	Values
Focal length (mm)	(f x, f y)	(4894.6412, 4812.1945)
Principal point (mm)	(u0, v0)	(1218.0733, 833.9089)
Distortion	(k1, k2, p1, p2)	(−0.060, 0.247, 0, 0)

Table 2. Interior parameters of the right camera.

Physical Meaning	Parameters	Values
Focal length (mm)	(f x, f y)	(4908.0802, 4795.6567)
Principal point (mm)	(u0, v0)	(1217.8019, 786.0223)
Distortion	(k1, k2, p1, p2)	(−0.062, 0.422, 0, 0)

The relative position relationship between binocular cameras can be obtained from the position matrix of the left and right cameras, denoting the rotation translation matrix from the right camera coordinate system to the left camera coordinate system. The conversion relationship among T_{LR}, T_L, T_r, R_{LR}, R_L, and R_R is expressed in Equation (9):

$$\begin{cases} T_{LR} = T_L - {R_R}^{-1} T_R \\ R_{LR} = R_L {R_R}^{-1} \end{cases} \tag{9}$$

The baseline distance, defined as the magnitude of the translation vector T plays a critical role in depth accuracy. According to the triangulation error model $\sigma_Z \approx Z^2 / fb\sigma_d$, where Z is the working distance, f the effective focal length, $B = ||T||$ the baseline, and σ_d the sub-pixel localization error of the laser stripe center; a larger baseline improves depth precision. However, an excessively large baseline reduces the overlapping field of view and may cause occlusion of reflective surfaces.

3. Principle of 3D Measurement System

3.1. Light Stripe Center Extraction

Laser triangulation is a technique used in structured light measurement. It employs a line laser to project a linear beam onto the surface of the target object, generating a light stripe.

As the depth of the surface of the object being measured changes, the shape of the light stripe changes accordingly. By capturing the image of the light stripe with a camera and performing a series of processing using image processing software, the three-dimensional

https://doi.org/10.3390/photonics13030238

coordinates of the points on the surface of the object under test can be obtained. In this way, the 3D reconstruction of the surface of the measured object can be realized.

The laser stripe is the only source of information for the line-structured light 3D measurement system, so the accurate extraction of the laser stripe center is directly related to the accuracy of the system. In view of the accuracy problem of the existing algorithms in extracting the laser center of the stripe, this paper adopts the Steger algorithm [33] based on the Hessian matrix, which can realize the sub-pixel accuracy positioning of the center of the light stripe: the specific steps are as follows:

Step 1: Image filtering process.

The image is reprocessed by convolution using the smoothing kernel of the second-order derivative of the Gaussian function. To reduce the influence of noise on the extraction accuracy, we set the Gaussian standard deviation as $\sigma = \omega/2$, where ω is the average width (in pixels) of the laser stripe in the captured image. In our implementation, the laser stripe typically has a width of $\omega \approx 3$ pixels, yielding $\sigma = 1.5$. Accordingly, the Gaussian smoothing kernel is implemented with a size of 9×9 pixels. This kernel size ensures that over 99% of the Gaussian energy is retained while effectively suppressing high-frequency noise.

Step 2: Hessian matrix representation of pixels

To solve for the normal direction of each pixel in the striped region in the Hessian matrix, the Hessian matrix for any pixel (x, y) can be expressed as the following Equation (10).

$$H(I(x,y)) = \begin{bmatrix} \frac{\partial^2 I_H}{\partial x^2} & \frac{\partial^2 I_H}{\partial x \partial y} \\ \frac{\partial^2 I_H}{\partial x \partial y} & \frac{\partial^2 I_H}{\partial y^2} \end{bmatrix} = \begin{bmatrix} r_{xx} & r_{xy} \\ r_{xy} & r_{yy} \end{bmatrix} \tag{10}$$

In Equation(10), $H(I(x,y))$ denotes the Hessian matrix of each pixel in the striped region. r_{xx},r_{yy},r_{xy} denote the second-order partial derivatives.

Step 3: Calculate sub-pixel coordinates

In the Hessian matrix, the eigenvalue with the largest absolute value represents the second-order derivative of the grayscale image. Its eigenvector determines the normal direction of the light stripe, denoted by (n_x, n_y). The sub-pixel coordinates can then be obtained using Taylor expansion in the normal direction. By solving the Hessian matrix at the point (x_0, y_0) as the reference point, the sub-pixel coordinates of the center of the light stripe are given by Equation (11):

$$(x_s, y_s) = (x_0 + tn_x, y_0 + tn_y) \tag{11}$$

t in Equation (12) is

$$t = \frac{n_x r_x + n_y r_y}{n_x^2 r_{xx} + 2n_x n_y r_{xy} + n_y^2 r_{yy}} \tag{12}$$

When $(tn_x, tn_y) \in [-0.5, 0.5] \times [-0.5, 0.5]$, it can be determined that the first derivative equal to zero is within the pixel being operated. Also, if the second-order directional derivative corresponding to the normal direction (n_x, n_y) is greater than at a given threshold, the point (x_0, y_0) is the center of the light bar and (x_s, y_s) are the sub-pixel coordinates. The effect of stripe extraction obtained using this algorithm is shown in Figure 3a,b:

https://doi.org/10.3390/photonics13030238

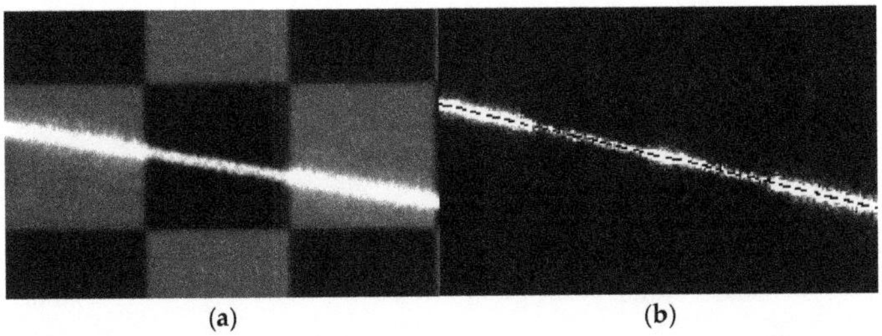

Figure 3. (**a**) Pre-processed light stripe, (**b**) light stripe center extraction.

3.2. Constructing 3D Models and Data Fusion

In this paper, a rotating platform is used to rotate and scan the reconstructed object. The Stokes vector formalism [34] is used to model the polarized light intensity distribution in the captured image.

The Stokes vector can represent various polarization states [35]. The expression form of the Stokes vector for polarized light states is mainly composed of four vectors, and the formula can be expressed as Equation (13):

$$S = [S_0, S_1, S_2, S_3]^T \tag{13}$$

In Equation (11), S_0 represents the overall intensity of this light wave, and S_1, S_2, S_3 represents the polarization state under three different angles, independent of each other. Due to the low probability of the occurrence of circularly polarized light, we ignore S_3 in this paper, and assuming $S_3 = 0$, the expressed formula for Equation (14) is as follows:

$$S = \begin{pmatrix} S_0 \\ S_1 \\ S_2 \end{pmatrix} = \begin{pmatrix} I(0^\circ) + I(90^\circ) \\ I(0^\circ) - I(90^\circ) \\ 2I(45^\circ) - (I(0) + I(90^\circ)) \end{pmatrix} \tag{14}$$

The intensity of polarized radiation at the image pixel location (x, y), in the direction of the polarization angle θ is $I(x, y, \theta)DI$, expressed as Equation (15):

$$I(x,y,\theta) = \frac{1}{2}[S_0(x,y) + S_1(x,y)\cos(2\theta) + S_2(x,y)\sin(2\theta)] \tag{15}$$

Prior to scanning, the polarizing lens is rotated to capture images at polarization angles of 0°, 45°, and 90°. Let $I(\theta)$ denote the image intensity (i.e., pixel brightness) observed when the polarizer is set to angle θ. The optimal polarization angle, which minimizes specular highlights by selecting the orientation with the lowest intensity, is then determined as expressed in Equation (16):

$$\theta_{optimal} = \arg\min|I(\theta)| \tag{16}$$

Sequence images of the left camera, consisting of the left polarization camera and the line laser, and the right camera, consisting of the right polarization camera and the line laser at the optimal polarization angle, are acquired, respectively. After that, the acquired sequence images are processed to obtain the laser region. The laser center is derived using Steger's algorithm for the laser region, using the interior parameters of the respective system camera and the optical plane equation. The 3D coordinates of each laser center on each frame were solved separately, and the laser spatial 3D coordinate data based on the world coordinate system under the camera coordinate system were obtained for each frame of the respective image. Finally, the stitching of the point cloud data is completed.

https://doi.org/10.3390/photonics13030238

When data stitching is performed, data stitching in global mode is used. The left camera or right camera world coordinate system is used as the world coordinate system. When processing the sequence image, the rotation and translation matrix from the camera coordinate system to the world coordinate system is obtained. The specific stitching process is as follows.

Step 1: Extract the laser center of the 1st frame image and calculate its spatial 3D coordinates by using the principle of laser triangulation.

Step 2: Extract the laser stripe center from the ith frame image ($i = 1, 2, \ldots, 320$) and compute its corresponding 3D spatial coordinates. These coordinates are then transformed into the base coordinate system using the associated rotation and translation matrices, and subsequently matched with the 3D point cloud data defined in that system.

Step 3: Determine whether it is the last frame; if so, output the 3D point cloud result; if not, return to the second step.

After processing the left camera and the right camera, respectively, with the above steps, the 3D point cloud model of each of the two systems can be obtained. To obtain a 3D model with richer point cloud data, the spatial 3D data of the objects obtained by the left camera and the right camera are fused and unified. In the fusion process, the coordinate system of the left camera is used as the world coordinate system, and the 3D coordinate data under the corresponding camera coordinate system of the right camera are unified into the world coordinate system through the rotation and translation matrix of the relative position relationship between the two cameras. Then, using the rotation translation matrix of the left camera, all the 3D camera coordinate points are transformed into the world coordinate system. Finally, the point cloud data fusion is performed to obtain a 3D point cloud model of the object surface with richer data. The overall process is shown in Figure 4.

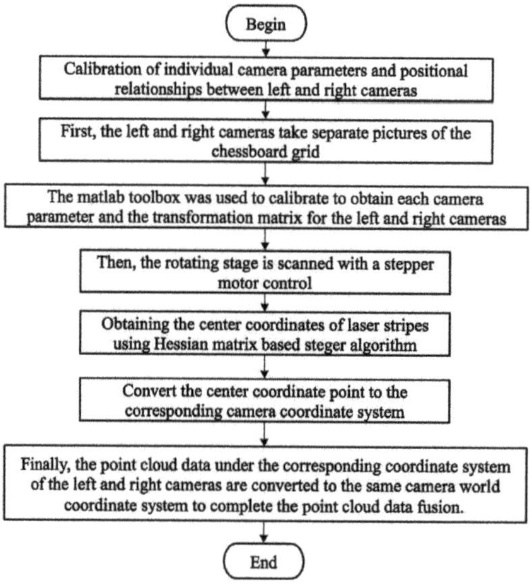

Figure 4. Calibration and point cloud data fusion process.

4. Experimentation

The 3D measurement system in this paper consists of two CCD cameras, two polarizers, a line laser transmitter, a rotating platform driven by a stepper motor for controlled object rotation (the stepper motor is used to rotate the platform), a computer, and a stand. As shown in Figure 5 below, a line laser is added to the binocular measurement system so

 https://doi.org/10.3390/photonics13030238

that the left and right cameras form a triangulation system with the line laser, respectively. Two Daheng image cameras, model MER-132-43U3C (resolution 1920 × 1200 pixels), Each camera is equipped with a Sony IMX267 global-shutter CMOS sensor(snoy Corporaion, Tokyo, Japan), featuring a pixel size of 3.45 μm × 3.45 μm, resulting in an active image area of approximately 6.62 mm (horizontal) × 4.14 mm (vertical)—corresponding to a standard 2/3-inch optical format. Consequently, the principal point coordinates reported in Tables 1 and 2 (e.g., $u_0 = 1219.073, v_0 = 833.9089$ for the left camera) are not exactly at the image center due to slight lens misalignment or non-uniform calibration coverage. However, they remain consistent across multiple calibrations and are used directly in the triangulation process without further correction, as their values are already optimized during the calibration procedure.

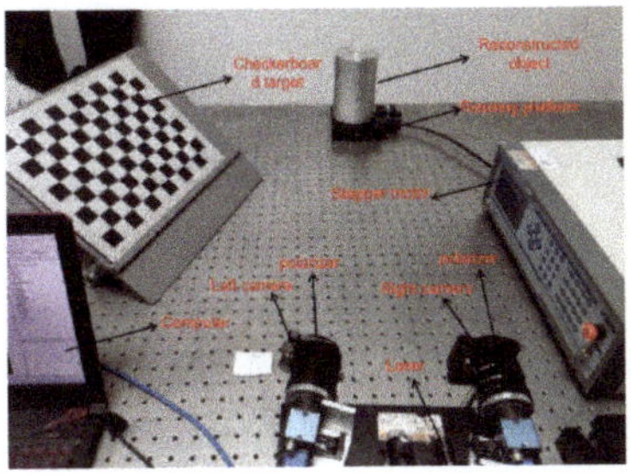

Figure 5. Binocular line-structured light vision sensor and reconstructed object in physics experiments.

Although these cameras are cost-effective, their intrinsic parameters may exhibit slight drift over time due to thermal or mechanical effects. To mitigate this, all measurements were conducted within two hours after calibration under stable laboratory conditions, and the system was recalibrated before each experimental session. Repeatability tests using a calibrated gauge block confirmed a 3D reconstruction stability of ±0.05 mm, indicating that parameter drift did not significantly compromise accuracy.

The illumination source consists of red line lasers with a wavelength of 650 nm. Both cameras and lasers were rigidly mounted on an optical breadboard to ensure mechanical stability, while the target object was placed on a computer-controlled rotating platform. The experiment was divided into three parts: (1) stereo camera calibration and laser plane parameter estimation; (2) monocular 3D reconstruction based on polarized line-structured light; and (3) binocular fusion of complementary reconstructions, followed by quantitative analysis.

4.1. Camera Calibration

As mentioned in the previous section, the interior parameters of the cameras in the line-structured light vision sensor were calibrated using a camera calibration method based on Zhang's method for the optimal polarization angle. A mutually orthogonal checkerboard grid calibration plate is used to calibrate the cameras, and two cameras are fixed on a stable mount. A polarizer with an adjusted polarization angle is placed in front of the camera, and it is ensured that it captures the entire calibration plate. Take 15 images containing the calibration board using different angles and positions. Make sure to cover the entire field of view and try to cover as many different bit positions and distances as possible. Each plane consists of 108 square grids of size 25 mm × 25 mm. The number of corner points detected on the target is much larger than the number of corner points required for calibration, which can effectively improve the accuracy of parameter calibration. The calibration results for the left and right cameras are shown in Tables 1 and 2.

https://doi.org/10.3390/photonics13030238

Reprojection error refers to the pixel-wise discrepancy between the observed image coordinates of a calibration feature (e.g., a checkerboard corner or laser stripe center) and the coordinates obtained by projecting its corresponding 3D world point back into the image plane using the estimated camera parameters. It is commonly used as a metric to evaluate the accuracy of camera calibration: a lower reprojection error indicates better consistency between the geometric imaging model and the actual measurements. In this work, the reprojection errors of the two cameras obtained from the calibration are shown in Figure 6a,b, and the average reprojection errors of the left and right cameras are 0.06 and 0.05 pixels, respectively. The calibration results have a good effect on the accuracy of corner point extraction, and the captured calibrated images have high accuracy.

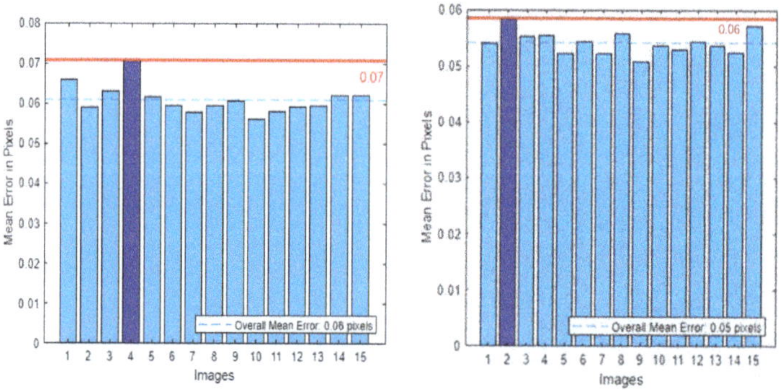

(a) (b)

Figure 6. (**a**) Reprojection error of the left camera, (**b**) reprojection error of the right camera. Red lines denote the maximum allowable errors (0.06 and 0.07 pixels). All errors fall below these thresholds, confirming accurate corner detection and high-quality calibration.

The conversion matrix between the two cameras is solved according to the principle of the second part, and the results obtained are shown in Table 3.

Table 3. External parameters between left and right cameras.

Conversion Relation	Rotation Matrix	Translation Matrix (mm)
Camera Left and Camera Right	$R_{Lr} = \begin{bmatrix} 0.976862 & -0.189362 & -0.099416 \\ 0.078374 & 0.749447 & -0.657409 \\ 0.1989954 & 0.634407 & 0.746947 \end{bmatrix}$	$T_{Lr} = \begin{bmatrix} -213.856038 \\ 996.732429 \\ 413.718675 \end{bmatrix}$

4.2. Light Plane Calibration

In this paper, we use Steger's algorithm with sub-pixel level extraction accuracy to extract the center point of the light stripe. The laser plane equation, where the center point of the light stripe is located, is fitted with the least squares method based on the extracted coordinates of the center point of the light stripe. The accuracy of the extraction results is evaluated using the root-mean-square error of the distance from the point cloud to the light plane equation. The evaluation parameters are listed in Table 4, and the calibrated results are shown in Figure 7a,b.

https://doi.org/10.3390/photonics13030238

Table 4. Line-structured evaluation parameters.

	Light Plane Equation (Z in mm)	RMSE (mm)	Coefficient of Determination
Left camera	$Z = 954.519 + 0.096 \times X + -1.067 \times Y$	0.0660	0.899
Right camera	$Z = 974.982 + -0.014 \times X + -1.077 \times Y$	0.0705	0.910

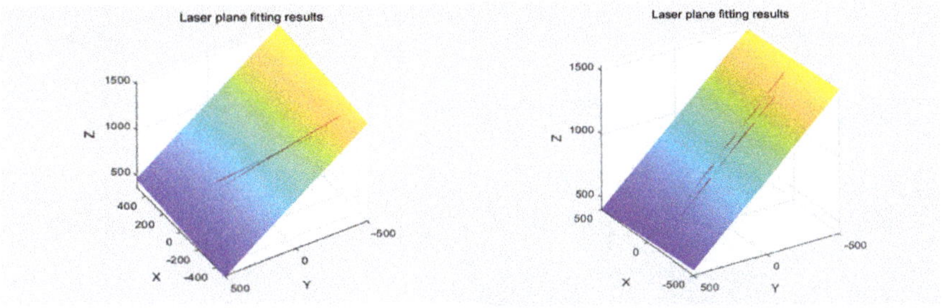

(**a**) (**b**)

Figure 7. Fitted light planes (**a**) left camera line structure light plane (**b**) right camera line structure light plane. The red lines represent the 3D spatial trajectories of the extracted laser stripes used for fitting. The colored surface indicates the best-fit laser plane, where the color gradient (from blue to yellow) corresponds to the values of the Z-axis coordinates.

As can be seen from the intensity of polarized 4, the RMSE obtained for the left camera and the right camera are 0.0660 and 0.0705, and the coefficients of determination are 0.899 and 0.910. Ideally the feature point on the optical plane is at a distance of 0 from the optical plane, and the RMES indicates the error in the distance between the feature point and the optical plane, and the closer it is to 0, the smaller the error is. The coefficient of determination indicates the fitting degree of the optical plane, and the closer it is to 1, the higher the fitting degree of the optical plane. From the calibration results, the accuracy of the calibration results is acceptable.

4.3. Monocular 3D Reconstruction and Binocular Fusion Results

Figure 8 shows an image of one of the left and right camera frames taken without a polarizer on a highly reflective cylinder, with very badly exposed light stripes (highlighted in red). The scanning reconstruction does not extract enough critical information, and there is a lot of noise. Using the method mentioned in the third part of this paper, the scanning reconstruction is performed on it based on the optimal polarization angle, as shown in Figure 9. As can be seen from the image, the overexposure of light streaks that occurs when imaging highly reflective subjects is effectively eliminated. The image quality is significantly improved(highlighted in red).

The Steger extraction algorithm is then performed on the image to extract the centers of the laser stripes. After that, the 3D coordinates of each laser center on each frame are solved using the interior parameters of the respective system cameras of the left and right cameras and the light plane equations, combined with the external parameters of the left and right cameras, as shown in Table 5:

https://doi.org/10.3390/photonics13030238

Table 5. Left and right cameras external parameters.

Camera	Rotation Matrix	Translation Matrix
Left camera	$R_L = \begin{bmatrix} 0.940484 & 0.235060 & -0.245433 \\ 0.016047 & 0.690679 & 0.722983 \\ 0.339460 & -0.683892 & 0.645801 \end{bmatrix}$	$T_L = \begin{bmatrix} -162.734252 \\ -99.261651 \\ 773.394442 \end{bmatrix}$
Right camera	$R_r = \begin{bmatrix} 0.993626 & 0.071979 & -0.086751 \\ 0.012412 & 0.695046 & 0.718858 \\ 0.112038 & -0.715353 & 0.689725 \end{bmatrix}$	$T_r = \begin{bmatrix} -141.477912 \\ -116.620106 \\ 796.625215 \end{bmatrix}$

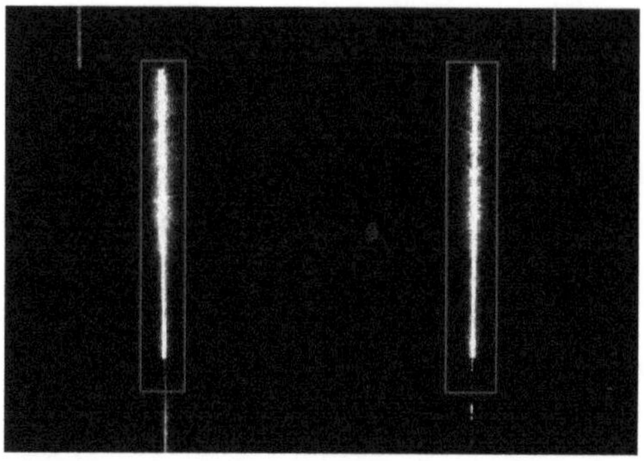

Figure 8. Images of a frame of the left and right cameras taken without polarizer on a highly reflective cylinder.

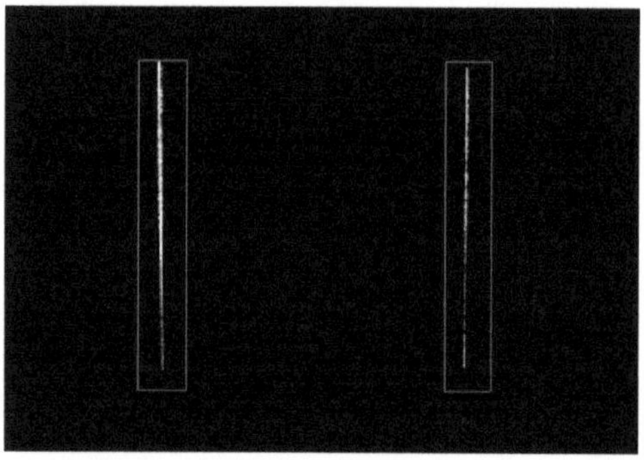

Figure 9. Images acquired by the left and right cameras during scanning reconstruction based on the optimal polarization angle.

According to these two sets of rotation and translation matrices, the laser spatial 3D coordinate data of each frame can be obtained based on the world coordinate system under the respective camera coordinate system. The complete spatial 3D model of the object can be obtained by point cloud stitching. The final visualized point cloud models of the left camera and the right camera are shown in Figure 10.

 https://doi.org/10.3390/photonics13030238

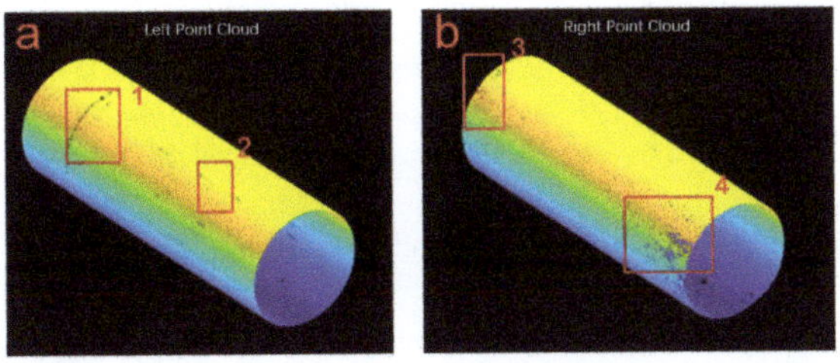

Figure 10. (**a**) Visual point cloud model of the left camera. (**b**) Visual point cloud model of the right camera. The red rectangles (labeled 1–4) highlight the reconstruction defects, including noise and missing data points, that occur when using a single camera view. The colored surface indicates the best-fit laser plane, where the color gradient (from blue to yellow) corresponds to the values of the Z-axis coordinates.

According to the experimental results of the reconstruction in Figure 10, we can learn that when the 3D reconstruction of highly reflective objects is performed for a single system, the point cloud information of the reconstructed object model is missing in some places (circled by red boxes 1, 2, 3, and 4 in Figures 10 and 11)

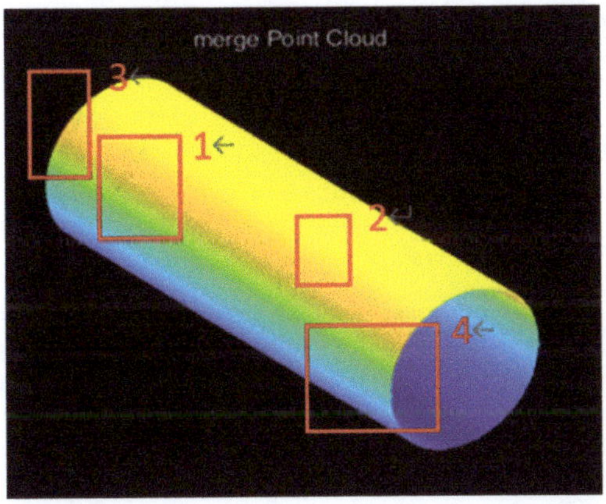

Figure 11. Visual point cloud model of the left and right camera. The red rectangles (1–4) correspond to the same regions shown in Figure 10a,b, demonstrating that the surface defects and missing data have been successfully filled and reconstructed through the multi-view data fusion process. The colored surface indicates the best-fit laser plane, where the color gradient (from blue to yellow) corresponds to the values of the Z-axis coordinates.

Afterward, in order to improve the accuracy and obtain more point cloud information, we will perform data fusion, and the experimental results are shown in Figure 11. The fusion result compensates well for the missing point cloud information of the single system so that the reconstruction results are more accurate and the 3D shape of the object is recovered completely. The experimental results show that the method is feasible and has high accuracy and robustness. It provides an effective reference for subsequent research work.

To further validate the feasibility of our method, we conducted experiments on standard spheres to discuss the analysis of the spherical light stripe center extraction results in an environment with or without light and optimal polarization angle.

According to the above two experiments, Figures 12 and 13 captured images and light stripes center extraction maps show that without the effect of polarization, the captured blob stripes maps have more severe highlights, and different lighting environments will

https://doi.org/10.3390/photonics13030238

significantly affect the light stripes extraction accuracy. The light stripes extracted with illumination show more noise, interruptions, and undulations than those extracted without illumination, missing a lot of information and many details. In order to improve the extraction of more information, the brightness of the light can be reduced to improve the extraction accuracy. After the introduction of the optimal polarization angle, the highlights are not limited to improved effectively but also extend to the noise, intermittency, and undulation phenomena occurring in the extraction, which are basically absent under different lighting conditions. Moreover, with the pre-processing in the light condition, the extraction effect is not much different from that in the absence of light. The results show that with the introduction of the optimal polarization angle, we can better extract more surface details from the object surface, provide richer information for the subsequent 3D reconstruction, and improve the accuracy of 3D reconstruction, and improve the accuracy of 3D reconstruction (Figure 14).

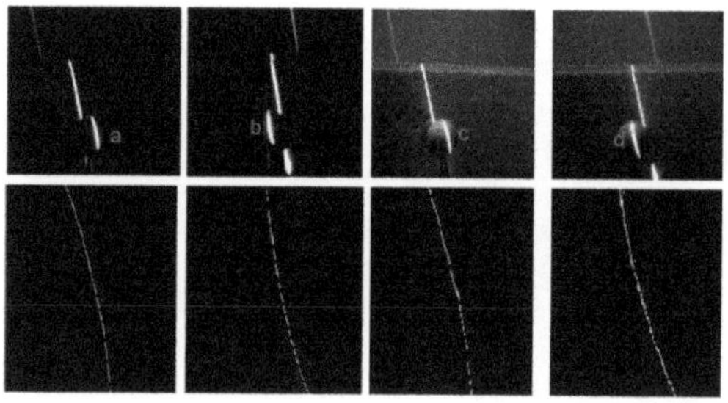

Figure 12. Pictures taken and extraction maps of the center of the light stripes. (**a**) Extracted image of the selected region of the blob stripe taken by the left camera in the absence of light. (**b**) Extracted image of the selected region of the blob stripe taken by the right camera in the absence of light. (**c**) Extracted image of the selected region of the blob stripe taken by the left camera in the presence of light, and (**d**) extracted image of the selected region of the blob stripe taken by the right camera in the presence of light.

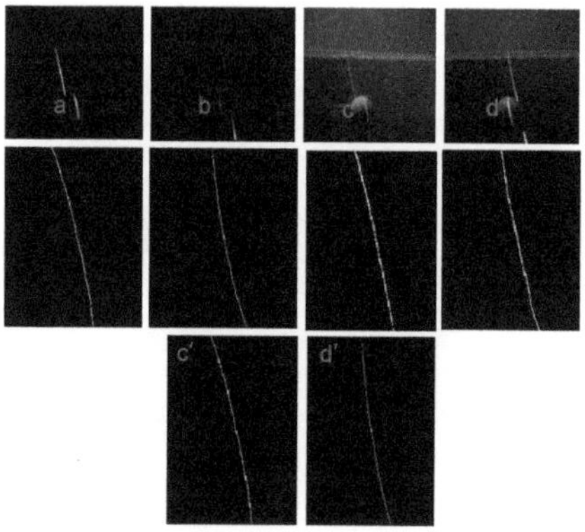

Figure 13. The image was taken based on the optimal polarization angle and the extraction of the light stripe center in the selected region. (**a**) The selected extraction region and the light stripe center extraction diagram of the image taken by the left camera in the light environment. (**b**) The selected extraction region and the light stripe center extraction diagram of the image taken by the right camera in the light environment. (**c**) The selected extraction region and the light stripe center extraction diagram of the image shot by the left camera in a light environment. (**d**) The selected extraction region and the light stripe center extraction diagram of the image shot by the right camera in a light environment. (**c′**) The pre-processed selected region extraction diagram under the condition of (**c**). (**d′**) The pre-processed selected region extraction diagram under the condition of (**d**).

 https://doi.org/10.3390/photonics13030238

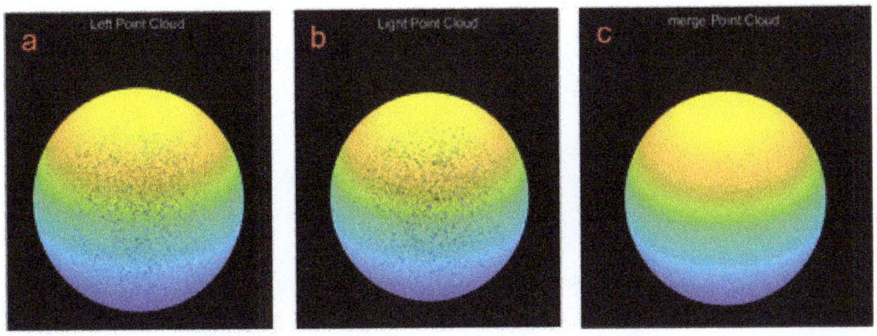

Figure 14. Diagram of the 3D reconstructed visualization model of a standard blob. (**a**) Visualization model of the left camera. (**b**) Visualization model of the right camera. (**c**) Fused visualization model. The colored surface indicates the best-fit laser plane, where the color gradient (from blue to yellow) corresponds to the values of the Z-axis coordinates.

The 3D reconstruction results show that the reconstructed blob produces information loss in random places, depending on the quality of the picture taken and the accuracy of extracting the center of the light stripes. Since the two systems take pictures and process information from different angles, the results of their respective reconstructions have similarities and differences. We fused the two model data by leaving the different point cloud data and removing the duplicated point cloud data, and finally, obtained our final 3D reconstructed model. However, as some information is still completely lost during the fusion process, for future improvement, we can continuously optimize and enhance the algorithmic model to achieve more accurate and reliable measurement results.

4.4. Comprehensive Evaluation: Robustness and Comparison with Alternative Methods

As shown in Section 4.3, the proposed polarized binocular method significantly improves reconstruction completeness compared to the non-polarized baseline.

To further evaluate robustness across surface reflectivity, we tested a polished metallic cylinder with machined features, both diffuse machined surfaces and highly specular curved regions. This represents a common industrial scenario where conventional structured light fails due to glare.

Figure 15 compares the reconstruction results without and with polarization. The non-polarized method exhibits large missing regions in specular zones (highlighted in red), while our method successfully recovers these areas. Quantitatively, point cloud completeness increases from 76.5% to 91.2%, with an RMSE of 0.09 mm relative to a reference scan.

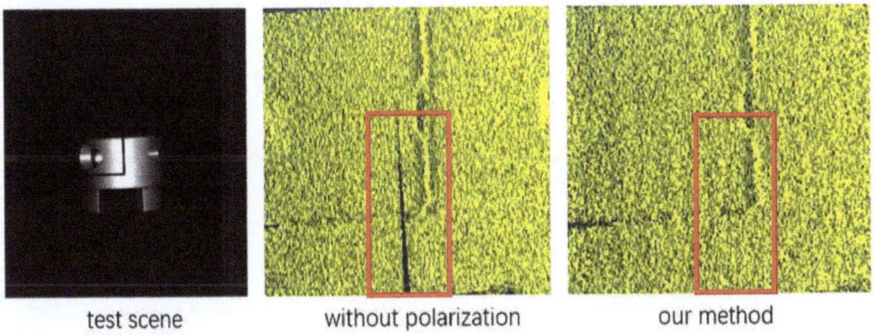

Figure 15. Reconstruction of surfaces with different roughness levels under non-polarized and polarized conditions.

https://doi.org/10.3390/photonics13030238

We also compare our method with a high dynamic range (HDR) structured light approach, which captures three exposures per camera (short, medium, long) and fuses them into an HDR image before stripe extraction.

As shown in Figure 16, although HDR reduces overall saturation, it still fails to recover stripes in near-mirror reflection zones—where all exposures are either over- or undersaturated. In contrast, our single-shot polarized method suppresses glare optically, achieving higher completeness (91.2% vs. 78.5%) and avoiding motion artifacts from multi-exposure capture. This makes it more suitable for real-time industrial inspection.

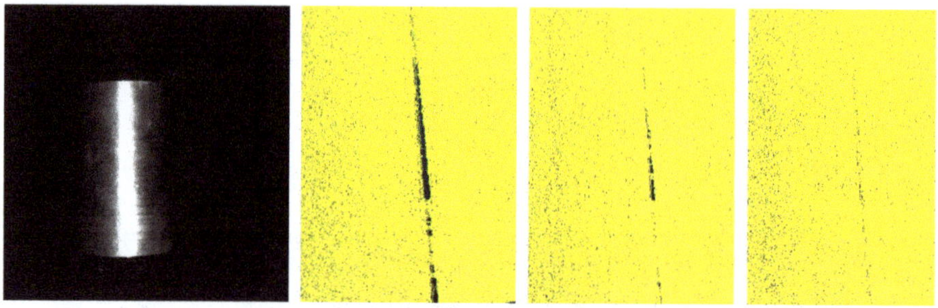

Figure 16. Comparison with HDR-based methods.

To further quantify the improvement in stripe extraction fidelity enabled by polarization, we analyze the light stripe images captured on a reflective cylindrical surface (Figures 8 and 9). For each image column, a Gaussian function is fitted to the intensity profile, and metrics including valid column ratio ($R^2 > 0.9$), centerline fitting residual, and stripe width (FWHM) are computed. As shown in Table 6, polarization increases the valid data coverage from 63.2% to 94.8%, reduces localization error from 0.37 to 0.11 pixels, and narrows the stripe width from 8.6 to 3.4 pixels. This demonstrates that polarization not only recovers missing stripes but also enhances sub-pixel extraction accuracy—providing a solid foundation for high-fidelity 3D reconstruction.

Table 6. Quantitative comparison of light stripe quality with and without polarization.

Metric	Without Polarization	With Polarization
Valid Column Ratio (%)	63.2	94.8
Centerline Fitting Residual (pixels)	0.37	0.11
Average Stripe Width (FWHM, pixels)	8.6	3.4

After binocular fusion, to eliminate redundant points caused by view overlap or registration errors, this paper employs a voxel-based downsampling approach. Specifically, the fused point cloud is partitioned into a uniform voxel grid with a certain resolution, and only one representative point—such as the centroid or the first sampled point within each non-empty voxel—is retained. This strategy not only effectively removes nearly duplicate points but also ensures a spatially uniform distribution of the point cloud while preserving geometric fidelity.

To evaluate the accuracy of the 3D models generated through data fusion, two validation methods are adopted: (1) assessing the number of points in the point clouds, and (2) comparing inter-point distances in the reconstructed models against ground-truth mea-

https://doi.org/10.3390/photonics13030238

surements. The point cloud data from 3D models obtained by different systems were comparatively analyzed using MATLAB R2022a, and the results are presented in Table 7.

Table 7. Comparison of the number of cylindrical 3D point clouds.

	Left Camera	Right Camera	Binocular Fusion System	Removing Redundant Points
Number of point clouds	322,637	309,901	632,538	557,549

From the table, it can be clearly seen that the binocular fusion system makes up for the information in the occluded area of the left and right cameras, which fully demonstrates the advantages of the binocular fusion system compared with the monocular system for collecting 3D topographic data. The experimental results show that the point cloud data obtained by the left camera contains 322,637 points, the point cloud data obtained by the right camera contains 309,901 points, and the fusion of the two pieces of point cloud data can obtain 632,538 points. The point cloud data after removing redundancy is 557,549 points, and the point cloud data is 74,989 points. It can be seen that the binocular system improves the number of point clouds by a factor of 1.7 compared to the monocular system, and the completeness of the reconstruction is improved.

The 3D model obtained by data fusion compares the fused measurements with the actual values. Table 8 records the measured values of the height (H) and diameter (D) of the cylinder measured in the three experiments compared to the actual values.

Table 8. Comparison of cylindrical measured and actual values.

	First	Second	Third	Measurement Averages	Actual Value	Δd	Relative Error
D/mm	100.10	99.97	100.11	100.08	100.00	0.08	0.08%
H/mm	149.95	150.04	150.06	149.95	150.00	0.05	0.03%

From the measurement results, it can be seen that the average value of the diameter of the cylinder from the three experiments is 100.08 mm, and the average value of the height is 149.95 mm. The absolute error of the diameter and height of the cylinder is 0.0 m and 0.05 mm, respectively. The relative error is 0.08% and 0.03%, respectively. Compared with Sun et al.'s method [10], the maximum measurement error of their method is not more than 0.16 mm for standard blocks, whereas the measurement error of our method is not more than 0.11 mm for cylinders. The measurement error of the 3D measurement system is within 0.1 mm, and the relative error does not exceed 0.1%, and the accuracy meets the requirements for application in 3D industrial measurement.

5. Conclusions

This work demonstrates that polarization-assisted binocular line-structured light effectively addresses missing data in the 3D reconstruction of complex, highly reflective objects. By placing linear polarizers in front of both cameras, specular highlights are suppressed, enabling the reliable extraction of light stripes in regions where conventional methods fail. Through fusion of complementary 3D data from left and right views, our approach achieves a 1.7× increase in point cloud completeness—for example, from 53% to 91% on a Polished metallic cylinder with machined features—compared to non-polarized binocular structured light. The system delivers a reconstruction accuracy of

 https://doi.org/10.3390/photonics13030238

0.08–0.11 mm RMSE and short-term repeatability of ±0.05 mm, which satisfies typical precision requirements for industrial dimensional inspection (e.g., tolerance classes per ISO 10360-8 for optical scanners).

Limitations remain for near-perfect mirror surfaces (e.g., uncoated glass), where polarization alone cannot recover structure. Future work will explore adaptive polarization states and post-reconstruction data fusion strategies to further enhance robustness and coverage.

Author Contributions: Conceptualization, Z.Z.; Methodology, L.Z. and Z.Z.; Validation, Y.Z.; Investigation, Z.W.; Resources, H.Z. and Z.W.; Data curation, Y.J. and D.X.; Writing—original draft, Y.Z.; Writing—review & editing, S.N.; Supervision, H.Z.; Project administration, M.L. and Y.J.; Funding acquisition, M.L. All authors have read and agreed to the published version of the manuscript.

Funding: This research was funded by the Jiangxi Provincial Department of Transportation Science and Technology Project (grant numbers 2024YB001 and 2023H0037) and the Jiangxi Provincial Key R&D Program (grant numbers 20243BBG21012 and 20252BCG330002).

Data Availability Statement: Data underlying the results presented in this paper are not publicly available at this time but may be obtained from the authors upon reasonable request.

Conflicts of Interest: Authors Zhong Wu and Dahai Xu were employed by the company Jiangxi Fangxing Technology Co., Ltd. Author Ling Zhu was employed by the company Jiangxi Communications Design and Research Institute Co., Ltd. The remaining authors declare that the research was conducted in the absence of any commercial or financial relationships that could be construed as a potential conflict of interest.

References

1. Marr, D.; Poggio, T. Cooperative computation of stereo disparity. *Science* **1976**, *194*, 283–287. [CrossRef] [PubMed]
2. Lemmens, M. A survey of stereo matching techniques. *Int. Arch. Photogramm. Remote Sens.* **1988**, *27*, 11–23.
3. Kraus, K. *Photogrammetry: Geometry from Images and Laser Scans*, 2nd ed.; De Gruyter: Berlin, Germany, 2007.
4. Luhmann, T.; Robson, S.; Kyle, S.; Harley, I. *Close-Range Photogrammetry and 3D Imaging*, 2nd ed.; De Gruyter: Berlin, Germany, 2014.
5. McGlone, J.C. (Ed.) *Manual of Photogrammetry*, 6th ed.; American Society for Photogrammetry and Remote Sensing (ASPRS): Bethesda, MD, USA, 2013.
6. Li, Y.; Qian, J.; Feng, S.; Chen, Q.; Zuo, C. Deep-learning-enabled dual-frequency composite fringe projection profilometry for single-shot absolute 3D shape measurement. *Opto-Electron. Adv.* **2022**, *5*, 210021. [CrossRef]
7. Kahn, S. Reducing the gap between Augmented Reality and 3D modeling with real-time depth imaging. *Virtual Real.* **2013**, *17*, 111–123. [CrossRef]
8. Yin, S.; Ren, Y.; Zhu, J.; Yang, S.; Ye, S. Vision-Based Self-Calibration Method for Robotic Visual Inspection Systems. *Sensors* **2013**, *13*, 16565–16582. [CrossRef]
9. Lu, R.S.; Li, Y.F.; Yu, Q. On-line measurement of straightness of seamless steel pipe using machine vision technique. *Sens. Actuators A Phys.* **2001**, *94*, 95–101. [CrossRef]
10. Zhang, G.; Sun, J.; Chen, D.; Wang, Y. Flapping motion measurement of honeybee bilateral wings using four virtual structured-light sensors. *Sens. Actuators A Phys.* **2008**, *148*, 19–27. [CrossRef]
11. Yang, R.; Cheng, S.; Yang, W.; Chen, Y. Robust and Accurate Surface Measurement Using Structured Light. *IEEE Trans. Instrum. Meas.* **2008**, *57*, 1275–1280. [CrossRef]
12. Mei, Q.; Gao, J.; Lin, H. Structure light telecentric stereoscopic vision 3Dmeasurement system based on Scheimpflug condition. *Opt. Lasers Eng.* **2016**, *86*, 83–91. [CrossRef]
13. Yi, S.; He, Z.; Wang, Y.; Yang, S. A Research on 3D Reconstruction Based on Structured Light. *Electron. Technol.* **2017**, *46*, 118–120.
14. Wang, Y.; Zhou, L.; Li, Z.; Wang, J.; Wu, X.; Wang, X.; Hu, L. Method for three-dimensional reconstruction of dynamic stereo vision based on line structured light using global optimization. *Measurement* **2024**, *224*, 113900. [CrossRef]
15. Sun, Q.; Ren, Z.; Zhu, J.; Dai, W.; Wang, M.; Sun, M. A Three-Dimensional Structured Light Vision System by Using a Combination of Single-Line and Three-Line Lasers. *Sensors* **2022**, *23*, 13. [CrossRef] [PubMed]
16. Huang, X.; Bai, J.; Wang, K. Target enhanced 3D reconstruction based on polarization-coded structured light. *Opt. Express* **2017**, *25*, 1173–1184. [CrossRef] [PubMed]

https://doi.org/10.3390/photonics13030238

17. Faugeras, O. *Three-Dimensional Computer Vision: A Geometric Viewpoint*; MIT Press: Cambridge, MA, USA, 1993.
18. Hartley, R.; Zisserman, A. *Multiple View Geometry in Computer Vision*, 2nd ed.; Cambridge University Press: Cambridge, UK, 2004.
19. Salvi, J.; Pagès, J.; Batlle, J. Pattern codification strategies in structured light systems. *Pattern Recognit.* **2004**, *37*, 827–849. [CrossRef]
20. Geng, J. Structured-light 3D surface imaging: A tutorial. *Adv. Opt. Photonics* **2011**, *3*, 128–160. [CrossRef]
21. Hao, X.; Sun, C.; Li, W. Three-dimensional reconstruction technology of pavement based on structured light. *Comput. Eng. Des.* **2015**, *36*, 2303–2307.
22. Chen, S.; Wang, Z.; Xiao, G. A monocular structured light 3D scanning method for large work-pieces. *Laser Infrared* **2018**, *48*, 5.
23. Wang, D.; Yang, L.; Zhang, Z. Line structured light scanning technique based on single and dual eyes combination. *Comput. Eng. Des.* **2020**, *41*, 749–755.
24. Guo, X.Z.; Shi, Z.Y.; Yu, B.; Zhao, B.; Li, K.; Sun, Y. 3D measurement of gears based on a line structured light sensor. *Precis. Eng.* **2020**, *61*, 160–169. [CrossRef]
25. Zhang, Y.; Wang, Y.; Qian, Y.; Liu, S. Improved 3D reconstruction method based on the Scheimpflug principle. *Opt. Rev.* **2020**, *27*, 283–289. [CrossRef]
26. Hu, P.; Cheng, S.; Li, S.; Liu, J.; Peng, X. Three-dimensional shape measurement based on multi-line structured light integration. *Mech. Eng. Autom.* **2018**, 29–31.
27. Zhang, Z. A Flexible New Technique for Camera Calibration. *IEEE Trans. Pattern Anal. Mach. Intell.* **2000**, *22*, 1330–1334. [CrossRef]
28. Zhu, Z.; Wang, X.; Liu, Q.; Zhang, F. Camera calibration method based on optimal polarization angle. *Opt. Lasers Eng.* **2019**, *112*, 128–135. [CrossRef]
29. Bouguet, J.Y. Camera Calibration Toolbox for MATLAB. 2013. Available online: https://data.caltech.edu/records/jx9cx-fdh55 (accessed on 25 September 2025).
30. Dewar, R. Self-generated targets for spatial calibration of structured light optical sectioning sensors with respect to an external coordinate system. In Proceedings of the Society of Manufacturing Engineers SME Vision '88 Conference Proceedings, Detroit, MI, USA, 9–11 June 1988.
31. Duan, F.; Liu, F.; Ye, S. A new calibration method for structural parameters of line-structured optical sensors. *J. Instrum.* **2000**, *15*, 1.
32. Li, D.; Wen, G.; Qiu, S. Cross-ratio–based line scan camera calibration using a planar pattern. *Opt. Eng.* **2016**, *55*, 14104. [CrossRef]
33. Steger, C. An Unbiased Detector of Curvilinear Structures. *IEEE Trans. Pattern Anal. Mach. Intell.* **1998**, *20*, 113–125. [CrossRef]
34. Goldstein, D.H. *Polarized Light*, 3rd ed.; CRC Press: Boca Raton, FL, USA, 2017.
35. Liu, J. *Research on Application Technologies of Binocular Stereo Vision Based on Polarization Imaging*; Xi'an Institute of Optics & Precision Mechanics; Chinese Academy of Sciences: Xi'an, China, 2021.

https://doi.org/10.3390/photonics13030238

MDPI AG
Grosspeteranlage 5
4052 Basel
Switzerland
Tel.: +41 61 683 77 34

Photonics Editorial Office
E-mail: photonics@mdpi.com
www.mdpi.com/journal/photonics

www.ingramcontent.com/pod-product-compliance
Lightning Source LLC
La Vergne TN
LVHW070447061126
851858LV00011B/177
9783725881239